CELEBRATE THE EARTH

BY DONALD IMSLAND

AUGSBURG PUBLISHING HOUSE
Minneapolis, Minnesota

CELEBRATE THE EARTH

Library of Congress Catalog Card No. 70-135219

International Standard Book No. 0-8066-9460-2

Manufactured in the United States of America

CONTENTS

To my wife Jan

for her encouragement and assistance

PREFACE

The countryside and the metropolis represent two extremes in the patterns of human settlement. While the one illustrates population sparcity, the other depicts an intense concentration of people in a relatively small land mass. To desire both and to love both may seem conflicting. But when they are viewed as two essential and interdependent parts of a total human environment, it becomes apparent that it is not "urban" or "rural" that one desires, but the Earth itself with its breathless beauty and myriad life forms. It is the magnetic appeal of the Earth which is the central theme of this book and which is at the heart of the present movement to save the environment. Whether farmer or urbanite, each has contributed to the present crisis and each has a common task: to rescue the Earth from a slow and final deterioration.

I grew up in a rural environment in the aftermath of the Great Depression. Patched overalls, sport shirts made from feed sacks, and kerosene lamps were parts of a commonly accepted style of life. Born in near-poverty, millions of my peers rose from "rags" to "riches" in one generation. It is apparent now that we progressed into an affluent society at a great and possibly fatal cost. Those who are alive at the turn of the next century will undoubtedly suffer from our engorgement of the Earth's riches. They will suffer because we embraced the myth that the Earth's environment was infinite. It may take generations to recover from this onslaught of human greed that insisted on brutalizing the Earth and consuming its goods to maintain a particular standard of living.

For many the threat of overpopulation and worldwide pollution came in 1970, which has been called the "Year of Environment." My awareness of the mounting crisis came early in the 1960s while I was a theological intern in Washington, D.C. In a class for engaged couples, I learned that certain people were choosing not to have children because of the rapid growth in the human population. At that time the problem did not seem as threatening as it does today. And the fact that individual couples were choosing childlessness out of a moral concern for the Earth made a deep impression on me. It began a slow transition in my perception of the religious community and its responsibilities to a world in serious trouble.

A second experience several years later reconfirmed my belief that population/environment problems must be attended to and soon, and that I personally must contribute toward their solution. Perhaps I was haunted by what later came to be a popular adage: "If you're not part of the solution, then you're part of the pollution."

I was attending the Aspen Seminar on Evironmental Arts and Sciences sponsored by the Thorne Ecological Foundation. Its purpose was to expose people from many disciplines to the young and increasingly popular science of ecology. Each morning small groups would accompany highly trained ecologists into the field to study the physical environment of the mountains. One noon we found ourselves eating sack lunches beside a small pond on a mountain plateau. Before sitting down we had to kick away the many frogs that lined the shore. As we sat eating our sandwiches, one member of the group remarked with alarm, "I just saw a frog eating another frog!" We looked in the direction he was pointing and saw the legs of a half-devoured frog protruding from the mouth of a larger frog. Almost simultaneously our group discovered that it was not a single incident. Frog cannibalism was taking place wherever we looked. Large frogs were eating medium-sized frogs, and medium-sized frogs were eating small frogs. Some were trying to devour frogs larger than themselves. It was happening with no emotion whatsoever—at least no apparent emotion.

The seminar leaders had no difficulty in transferring this experience to the afternoon discussion. We talked about the human population explosion now taking place on the surface of the Earth. We discussed the multination famine which may strike our planet with force in the mid-1970s.[1] We learned that the present rapid growth in population and the present concentration of humans in large urban centers has a direct parallel to the frog pond. There were too many frogs crowded into too small a land area and the food supply became depleted. Nature's solution was frog cannibalism.

The experience reminded me of the *Esquire* magazine article in 1967 entitled, "The Human Race Has, Maybe, Thirty-Five Years Left. After That, People Will Start Eating Plankton. Or People." It accentuated what many experts have been saying: mankind will soon reach the environmental limits on the surface of the Earth. Man will either control his numbers and stop his destruction of the Earth or turn his ancient home into a human frog pond.

It is no longer surprising that several well-known scientists have publicly expressed concern for the future lives of their children. In many respects the future looks dismal. Paul Ehrlich's threat of the "population bomb" has frightened every sector of society, and his book has been read by youngsters not yet in their teens.[2] John

Platt speaks of the "crisis of the crises" wherein a teachers' strike, a race strike, a garbage strike, and a police strike occur within a few days of each other. Platt believes that unless we find large scale solutions to multiple crises, we have less than a fifty-fifty chance of living until the end of the decade, and "anyone who feels more hopeful about getting past the nightmares of the 1970s has only to look beyond them to the monsters of pollution and population rising up in the 1980s and 1990s." [3]

The following chapters try to show that the environmental crisis is really a crisis of the Earth. When our planet is placed within the context of cosmic space we can only conclude that the spaceship on which we ride is indeed infinitesimal. When viewed from the perspective of geologic time, our benign habitat is ancient and mankind is but an infant. The thin, web-like structure of the Earth's surface represents millions of years of evolution and it is unlikely that an exact replica can be found elsewhere in the universe.

The present pollution of the environment is a threat to the basic equilibrium of life systems. While there are multiple causes, the underlying source of the problem appears to be man's attitude toward nature; this attitude is shaped by religious and cultural traditions. The Judeo-Christian religion has been cited as a primary cause since it extols man and depreciates nature. The evidence, however, does not bear this out since the biblical writers affirm the intrinsic goodness of the Earth and look forward to the ultimate reconciliation of "all things." We cannot, therefore, continue to propagate the "brotherhood of man" without at the same time placing equal emphasis on the "sisterhood of the biosphere."

It is important that we place a new emphasis on the future and learn to give the Earth a proper place in our lives, since we have grown out of its surface and are sustained daily by its resources. We celebrate and give honor to religious and national holidays, and we mark many occasions and events with ceremony and festivity. We must now learn to honor the Earth as a total system of life without which we could not exist. We must learn within the next few decades to remove once and for all the unfortunate dichotomy which separates man from nature and begin to "celebrate" the Earth as our friend and not our foe. In the religious community it is customary to "remember the Sabbath to keep it holy." Mankind must, in the remainder of this century, remember the Earth to keep open the possibilities for future life.

THIS BELEAGUERED EARTH

1

And I brought you into a plentiful land to enjoy its fruits and its good things. But when you came in you defiled my land, and made my heritage an abomination.

Jeremiah 2:7 RSV

For centuries man has indulged freely of the vast resources comprising his environment. Air, land, and water have been available in seemingly unlimited proportions. Space for movement, the beauty of landscape, and the tranquility of nature have been invaluable to the enrichment of human life.

Although he is often plagued with a rapid depletion of certain products, man has learned to apply massive forces toward their preservation. Only in recent decades has the regard for supply included a new concern for quality. Increasingly the demand is for *clean* air, *pure* water, and the *proper use* of land.

A panorama of new environmental problems has beset the mid-20th century. The Earth's exploding population and advancing technology are producing lack of space, water shortages, industrial wastes, noxious gases, offensive noise, urban blight, famine, and a host of other problems. These difficulties will neither go away soon nor easily. The crisis is *real, global,* and *growing*.

The quality of man's environment is a fundamental concern of many individuals and organizations. Increasing numbers have joined in the struggle to maintain the Earth as a suitable place for human life. They view the environmental crisis as a broad and interrelated set of complex problems which reflect *man,* who is at once the principle offender of the present condition and the chief creator of what is to come.

Symptoms of the Crisis

"Everyone wants to save the environment but no one knows quite what to do." [1] This statement depicts the perplexity which surrounds the environmental crisis. Concern for the environment falls easily into that special class called "motherhood" issues which no sincere citizen can afford to oppose. Everyone, from retirees to grade school children, is alarmed over the degradation of the Earth's natural systems. But few, if any, are able to grasp the breadth and depth of the multiple problems, and no one has been able to offer comprehensive and workable solutions.

The public response to the crisis in recent years is remarkable. Citizens have written letters to congressmen; they have joined crusading conservation groups; they have taken part in ecology rallies; and they have consumed hundreds of books and articles on the subject. New laws have been passed; billions of dollars have been spent on pollution abatement; and many new techniques have been devised to control the effluents of industry. But in spite of this great concern, the problems do not seem to go away—in fact, they multiply. The massive efforts directed toward the control of air pollution have had little effect on the general quality of the air shed over the United States. Similarly, the efforts which have been applied to cleaning up the lake and river systems have not substantially improved the quality of water.

Our progress, if we can call it that, has been to become ever more aware of the degree of widespread pollution. Increasing awareness causes us to want to save the environment, but few of us know what action will insure immediate and long-lasting results. There is an enigma surrounding the crisis. In part, it is due to our limited understanding of the problems, and in part it is our impotence to create permanent solutions. But beyond this, assuming that we achieve understanding and evolve solutions, Americans may be unwilling to make the necessary individual and collective sacrifices to overcome the destruction of the Earth. An ultimate solution may require a new set of priorities for our society, and a new life style for all citizens. But the perplexity still remains: how much will be expected of us, and are we willing to pay the price?

In his final report of 1965, former Secretary of the Interior, Stewart L. Udall, expressed concern that the sheer power of the population and technological revolutions may make man himself an endangered species in many parts of the world. Udall was referring to our intrusion on priceless habitats with dredges, bulldozers, pesticides, and concrete. He warned that we are destroying the places and food supplies which wildlife need for survival, and in destroying wildlife, we may inadvertantly be jeopardizing the future of our own species.

As early as 1947 a conference at Princeton University discussed the question of man's fate. Would he become extinct like the dinosaur, or would he assume responsibility for his own destiny? Opinion at that time was divided. But in 20 years the condition of man's environment has changed radically. There was general agreement at the UNESCO headquarters in Paris in late 1968 when more than 300 experts from 63 nations met to discuss the crisis in ecology. They decided that in 20 years life on the planet will be showing the first signs of succumbing to pollution: the atmosphere will become unbreathable for men and animals; life will cease in rivers and lakes; plants will wither from poisoning. The opinion was made public in a final report entitled, "Intergovernmental Conference of Experts on the Scientific Basis for the Rational Use and Conservation of the Resources of the Biosphere." [2]

Partly as a result of the Paris meeting, the United Nations General Assembly agreed unanimously in December 1968 to hold a Conference on the Human Environment in 1972. James Russell Wiggins, then U.S. Ambassador to the United Nations, stated that this will be the most momentous of all decisions made by the General Assembly in 1968. Wiggins noted:

> *We are dealing here with the life and death of the human family, and surely when the history of our age is written, and the records are studied a thousand years hence, the world will say of this decision of the General Assembly in 1968—it was about time.*[3]

Many have predicted that time is running out for man. Barry Commoner of Washington University told the American Association for the Advancement of Science in late 1969 that perhaps within the next generation a dual world-wide catastrophe—overpopulation and pollution of the environment—will confront mankind with the threat of extinction or with a gross deterioration in the level of society. Others believe that something more serious than extinction could develop. René Dubos, formerly of Rockefeller University, believes that man can adapt to almost anything, and he is not on his way to extinction. The real tragedy is that we *can* adapt to multiple forms of pollution so prevalent in the large cities. Dubos writes, "If we do not understand what the environment is doing to us, something worse than extinction will take place—a progressive degradation of the quality of human life." [4]

Environmental disturbances are taking place throughout the globe. The list of potentially damaging pollutants grows longer each day. We must find ways to effectively reduce the detrimental effects of air and water pollution.

But we must also learn to control the use of pesticides and fertilizers, radioactive wastes from nuclear power plants, oil spills from tankers and blowouts from offshore drilling, the rising carbon dioxide level of the atmosphere, noise pollution from jets, the disappearing plant and animal species, the continuing growth of the world's population, and the physical and psychological effects of urban concentration.

It has now been well established that Lake Erie is dying. The lake is growing old before its time. All lakes grow old as they collect runoff and materials from the surrounding shores. It is true of pristine lakes far removed from human settlement. Over thousands of years they eventually accumulate enough silt from erosion and organic materials to turn them into marshes, and finally into dry land. But in Lake Erie, man has greatly accelerated that process with his wastes and sewage. Because the lake is shallow, there is hope that the rate of pollution can be retarded through improved sewage treatment and proper oxygen circulation. But whether or not Erie can be saved is not certain. What is certain is that "in the past 50 years, Lake Erie has aged 15,000 years."[5]

The city of Los Angeles has become this nation's symbol of air pollution. It is deeply involved in a struggle to control smog. In several cases, school children in the Los Angeles area have been prevented from playing outside on days when the smog is most intense. The smog has also been responsible for seriously damaging millions of trees in various parts of California. Several years ago it was discovered that smog destroys the chlorophyll in the leaf tissue, and the aging process accelerates, weakening the trees and making them prey to forest insects, particularly bark beetles. The problem has grown worse year by year with increasing population and industrialization. But the main culprit appears to be the automobile. The state of California was the first to propose a law prohibiting the use of the internal combustion engine after 1975. This would require an alternative source of power to drive the family car. Although the proposal was defeated, it does indicate a growing concern among citizens of that state.

Los Angeles is not alone in the battle to control air pollution. Every major urban center has a problem. A crisis occurred in New York during the summer of 1970 when a temperature inversion held the noxious fumes captive over the city for several days. During the same period Minneapolis recorded its highest degree of air pollution in that city's history. Pollution is reportedly killing cattle in Bulgaria, defoliating trees in Yugoslavia, sending water-drinkers to the hospital in Hungary, and destroying the beautiful old baroque facades of Czechoslovakia. In a warning to the Greek government by their Chamber of Technology, it was reported that Athens, one of the oldest cities in Western civilization, may have to be abandoned in 10 years unless steps are taken to combat air pollution. Air pollution has become a global problem and there is every indication, in spite of the corrective measures being taken, that it will become worse. Meteorologist Morris Neiburger has warned:

> *If air pollution increases beyond the capacity of the atmosphere to cleanse itself, smog will encircle the earth, and all of civilization will pass away. Not from a sudden cataclysm, but from gradual suffocation by its own effluents.*[6]

Rachel Carson called ours the age of poisons and with good reason, for many of the thousands of new chemicals produced each year are not only highly toxic to living things, but amazingly persistent. DDT now is everywhere and remains toxic for years. Unlike many of the organic materials produced by living things, it is resistant to breakdown by microbes, water, and sunlight. Once it is released, it may travel considerable distances without losing its toxicity. It may be eaten and then excreted by most higher animals without suffering substantial change. Once distributed over the face of the globe, DDT may be picked up by a plant, passed on to an animal eating the plant, returned to the soil when the animal dies, cycling through the world of living things for decades. Several animals, including man, which are at the end of a long food chain are most likely to receive DDT in concentrated form. The pesticide is now present in rain and surface water in concentrations of several parts per trillion, but most people carry DDT in their fatty tissue at a level which is one million times higher. There are few foods in the human diet free of DDT. Recent studies show that even mothers' milk passes some DDT to infants.

DDT represents an "unforeseen, international, ecologic boomerang."[7] For years it has been used to rid communities of undesirable pests. The military used it in World War II to delouse prisoners. But its most widespread use is in the control of insects which damage crops such as cotton. In recent years the boll weevil has become increasingly immune to DDT and its damage to cotton crops in Southern states continues. Recently it has been reported that DDT has effectively killed off the natural predators of the weevil making the problem of control more complex. The ecological boomerang resulting from the massive use of DDT may be repeated in the widespread use of other insecticides and herbicides.

Beaches from England to Puerto Rico to California have been soaked in oily slime. Fish and wildlife have been destroyed. Clean, tranquil shorelines have been coated with oil sludge making them undesirable for humans and wildlife. The Santa Barbara oil slick created great alarm over the possible widespread destruction of beaches and marine life. The offshore blowout spewed more than 3,000,000 gallons of oil in 100 days of leakage. The embittered citizens of Santa Barbara wish to declare the Outer Continental Shelf Land Act unconstitutional on the basis that the right of people to free and unencumbered use of the environment is a civil right, as basic as the right of free speech.

Our government is spending billions of dollars to create the supersonic transport (SST) which promises to disturb thousands for every one person who is whisked from coast to coast three hours faster. The SST will be the noisiest aircraft ever flown. At supersonic speeds it promises to create a path of sonic booms 50 miles wide. Waiting for takeoff it will cause an ear-shattering sound that some experts have equated with the sideline noise of several dozen ordinary jets. Those promoting its development claim that by the time it is built the problem of ground noise will be solved. But there still remains the destructive potential of its sonic boom, plus the long-range effects of stratospheric pollution. The essential question is whether these hazards are worth the benefit of saving some travel time for a fraction of the world's population.

The proposed Alaska oil pipeline has been tentatively approved. The four-foot diameter pipeline would carry oil 800 miles from the North Slope of Alaska to a seaport on the southern shore of the state. Among the ecological consequences anticipated by the U.S. Geological Survey is the melting of the permanent frozen tundra along the route of the pipe which would have to be heated to keep the oil flowing in the cold climate. The thawing would result in mud flows along the path of the pipeline. Such mud flows would occur even in nearly flat basins. The effects of the mudslides would be an uncovering of and loss of support for some sections of the pipeline, and valleys and depressions would fill with mud.

We are rapidly passing from a phase of energy abundance to one of energy scarcity. Reports indicate that the natural gas supply will be under a severe strain until well beyond 1975. The decline of the domestic oil supply has been somewhat relieved by the ready availability of imported oils, but in many ways it has been even more severe than gas. Our domestic supply, so vital in the event of emergency, is declining steadily and likely will be gone within the next several years. At this point we shall be dependent upon foreign sources. We are shy on electrical generation as well. We are entering a period of growing scarcity in energy of all kinds, and the ironic thing is that we go on acting as if our supply is endless.

Climatologist Reid A. Bryson claims that the climate of the Earth has shown a cooling trend since 1940. He blames the downward trend of temperature on the increase of smoke and pollution in the overburdened atmosphere. Other scientists believe that the Earth's environment is gradually becoming warmer. One school holds that the increase in temperature will melt the icecaps of Greenland and the Antarctic raising the sea level by as much as 300 feet, thereby obliterating most of the major cities of the world. Another school believes that the high temperatures will bring about an increase in evaporation and with it a sharp rise in precipitation; the additional snow falling on the icecaps will start the glaciers moving again and another ice age will be upon us.

The problems seem insurmountable. We have not dealt adequately with the environmental issues of the 1960s. Now we are not only faced with their continuation, but we must somehow confront new problems emerging on the horizon. It seems clear that solving one of the problems is no real solution. There must be a comprehensive approach to the environmental complex if we are to move ahead at all. The important questions of mankind can be answered only through a cooperative effort involving both experts and citizens. The new "Jeremiahs" of the ecological movement are warning us that there is not much time left. They do not hesitate to predict the end of the world, or at least the end of life as we know it. Perhaps we are in a period of grace and there is yet time to save the environment if we know exactly what to do.

Searching for Causes

It is customary to blame our environmental ills on some abstract cause. Not uncommon is the attitude that "pollution is a product of our age." The popular rationale goes something like this: if we wish to maintain our present standard of living, we must be willing to accept a certain amount of waste, noise, and congestion. Because we are living in a highly technological society, which is itself a symbol of progress, we must learn to accept the bad with the good. In essence, advancing technology and rapid change are inevitable and to some extent necessary. Therefore, we must learn to "adjust" to these new conditions in order to progress to even higher living standards.

Our sophisticated technology has often been identified as the prime cause of environmental pollution. The knowledge of science, applied to the conversion of raw materials into consumer goods, has produced a steady flow of merchandise, much of which we do not need. The average American consumes several times more food than the average person from countries which contain two-thirds of the world's population. The average American is responsible for 2,500 pounds of waste per year—many times the world's average. The United States, with less than six percent of the world's people, accounts for almost 30 percent of poison being dumped into the sky and seas.

Technology can produce anything, and in mass. If the product can be made, and if there is a reasonable profit, technology will find a way. It has been the main contributor to the population explosion because it has found effective ways to decrease infant mortality. In the future it will greatly extend longevity through organ transplants and new drugs. Applied to the development of weaponry, technology has placed man in a position where he can destroy all of life several times over. Applied to the automobile, technology has created a situation where cities are now experiencing a "paralysis of movement" while every 24 hours more than 10,000 new automobiles are produced, and each year several million cars are discarded. Applied to air travel, technology has created a condition in the United States where an average of 10,000 planes are aloft over the continent at any given time during peak weather conditions, and projections indicate that this number will increase to 50,000 by the year 1990.

We are married to technology, and it is not likely that we will soon divorce ourselves from its shackles. The freedom of technology to invent whatever it happens to invent is taken as the underlying law of modern life. But technology is not an autonomous force. It is merely a tool for man. Human decisions determine what forms technology should take. If technology has been a detriment to the environment, we should not blame technology. We should rather question the motives of those who utilize technology in ways which are destructive of life. After determining that a product can be made, those in control of technology have traditionally failed to ask whether it will create negative ecological consequences. In the future we must enforce effective controls over technology so that ecological considerations will be given the same priority as economic factors.

Closely aligned with technology as a cause of pollution is the profit motive. We all have a high regard for a quality environment. We desire pure air, clean rivers for recreation, beautiful landscapes, and all the rest. But there are some who have an even greater passion for profits. These profits are too often generated by increased auto production, by oil deposits along coastal shores, or by the ruin of an entire valley by strip mining. Corporate interests have not wanted to pay the social cost of their private, money-making investments. Above all, they do not want to be told to stop what they are doing—not by conservationists and certainly not by state or federal governments. The priorities of oil companies, automakers, utilities, and others can be, and often are, profoundly anti-social, changing man's relationship with his environment. Builders, bankers, investors, and realtors hot on the scent of capital gains are tireless in their demands for "continuous growth" which in most cities means crowding, traffic congestion, urban sprawl, and all the scores of problems each of these creates. The statistical information regarding auto output, steel production, and energy needs have become the indicators of progress. Our society operates with an ideology of economic determinism. Our Holy Grail is the gross national product. And as a result, the quality of our natural environment is greatly reduced.

In addition to undisciplined technology and the profit motive, perhaps a more fundamental cause of the present environmental malaise is population density. It is not only a problem of population growth, but a problem of concentration—too many people living in too small a land area. When urban residents discover that public facilities such as schools, parks, transportation, and electrical power are grossly overloaded; when they feel oppressed by crowded streets and jammed buses; when they experience hostility and inhumanity among citizens; when they have trouble finding housing or a parking place, they probably would conclude that adding another 100 million people to this nation's population by the end of the century is not such a good idea. The growth and concentration of humans in urban centers appears to be inevitable and will continue. But these are direct causes of environmental decay, particularly when we are informed that the United States consumes more than 50 percent of the world's productive resources. Increasing population, increasing concentration, and increasing affluence in this country can only result in placing the entire world in jeopardy.

Not as obvious as a cause of pollution is the question of who owns and who is responsible for the "public" environment. Who owns the air shed over a large city? Who owns the river system which transverses a metropolitan area? Throughout its development the industrial system has assumed that it was proper to use rivers and air space as public cesspools. Municipalities have utilized waterways to carry away millions of tons of wastes. Large companies have obstructed the public view with massive billboards and gaudy signs. Airline companies bombard entire communities day and night with the rumble of their jets. Private citizens in many communities burn trash in backyard incinerators, paying no heed to its contribution to air pollution and the stench it creates for their neighbors. Individual motorists produce noise pollution with their unmuffled exhausts and blaring horns. We treat the public environment as a "no-man's-land." It belongs to no one and, therefore, it is left to the discretion of each individual how he will use it. Garrett Hardin describes this practice as "The Tragedy of the Commons." [8] The public environment is a "commons" belonging to everyone, and too often it becomes the receptacle for everyone's wastes.

The late Thomas Merton, a Trappist monk from Kentucky, believed that the ecologic crisis stems from the fact that we in America are operating on a "frontier mythology." [9] We continue to live today with attitudes like those of the Puritans who saw the land as a great wilderness frontier which they felt compelled to conquer and destroy. Merton argues that the Puritans had a half-conscious bias against the realm of nature. They hated it as a person, and the Bible gave them plenty of texts to justify their traditional repugnance toward the wilderness. This attitude toward nature did not die with the pioneers. It remains as a basic attitude today particularly within business and industry. We all proclaim our affection for wild nature, and in the same breath we affirm our strong attachment to values that demand the destruction of the last remnants of wilderness.

René Dubos observes that the first phase of American history was a "saga of conquest," the enemy being nature itself. The second phase was devoted to the exploitation of mineral resources and the creation of industries. When it had mastered the wilderness, the pioneering spirit set out to master all other manifestations of nature. Dubos writes:.

> *The American way of life, now spreading all over the world, is based on the myth that man is the Lord of Creation. This has become a credo and almost a religion of the world's masses and their leaders.*[10]

Lynn White Jr. believes that at the root of our ecological crisis lies a "Christian arrogance toward nature." He points out that before the Christian era trees, springs, hills, streams and other objects of nature had guardian spirits. These spirits had to be approached and placated before one could safely invade their territory. "By destroying pagan animism," states White, "Christianity made it possible to exploit nature in a mood of indifference to the feelings of natural objects."[11] White claims that it is probably in vain that so many look to science and technology to solve our present ecological crisis. We need to go to the essential cause which is religious, and in particular, to the values of the Judeo-Christian tradition.

We cannot avoid the statement in Genesis 1:28—"Be fruitful and multiply, and fill the earth and subdue it; and have dominion over the fish of the sea and over the birds of the air and over every living thing." A literal interpretation of this mandate has led to an overemphasis on man's superiority over nature. But this command, in the first version of the creation story, must be balanced with the second story in Genesis 2:15—"The Lord took the man and put him in the Garden of Eden to till it and keep it."

It is clear that the biblical writers place man in authority over the rest of creation. This is a unique privilege. But privilege does not come without responsibility. The command in the second story of creation makes it quite clear that man is responsible for the "garden" and the garden is the Earth. The biblical accounts will not condone the careless exploitation of nature which has gone on in this century. But the religious community has remained virtually mute on the problems of pollution until recently. And for this reason the accusation of Lynn White Jr. regarding Christian arrogance toward nature can be generally substantiated. The church and its institutions have not articulated an ethic dealing with man's relationship to his environment. It is obvious now that such an ethic will be as necessary for man's future as the social ethic which relates man to his brother.

Romanticized Solutions

The faith of the American people in technology is astounding. For the average citizen the promises of technology far outweigh its potential threats. A common complaint of the socially conscious citizen is: we have enough science and technology to place men on the moon and return them safely to the Earth, but we cannot maintain a decent environment in our land.

The implication is clear: the solution to our ecological crisis is technical. The way to solve the problems of environmental pollution is simply to apply more technology. This reflects the dominant American philosophy, and it suggests that "anti-pollution" may become the next giant industry. Certainly we must develop many new technologies to combat pollution, but we must guard against the possibility that the next era of technology will produce even greater environmental hazards. We must not too readily assume that the piecemeal, short-term solutions of technology are the solutions we really desire.

It has been well established that the automobile is the greatest single contributor to air pollution. Therefore, we must exercise some kind of control. As an initial step the auto industry has designed an exhaust device which must be placed on all new models. Although the mechanism does not measure up to its promises, the attitude of the industry is that the problem will be dealt with if it is given sufficient time. In the meantime, the federal government has fostered an effort to find an alternative to the internal combustion engine. The possibilities are steam engines, gas turbines, electric cars, and other exotic energy conversion techniques.

While the internal combustion engine is being either improved or replaced, it continues to spew its noxious fumes. But the fallacy is to assume that replacement of the internal combustion engine will make the automobile safe for the environment. This solution does not confront the growth of highways which cover millions of acres of oxygen-producing greenery each year. And it does not solve the problems of congestion, traffic snarls, service stations, parking lots, and the growing incompatability of the auto with city life.

In addition to more and better technology, a far greater allocation of private and public funds must be applied to the solution of environmental problems. Federal, state, and local governments have begun to respond to the enormous cost of pollution control. Business and industry have been forced to direct larger sums toward pollution abatement. But the dedication of government to solving the problems of pollution must be seen in relation to other priorities. The present administration has spent more than any other on environmental protection. However, compared to other items on the national budget, it is a pitifully small amount in relation to what some have estimated the cost to be. Likewise, most major industries throughout the country have stressed that they are spending millions of dollars to install pollution control devices. In relation to what has been spent in past years this is impressive. But when compared to the amount of annual profits, the expenditure does not represent a significant sacrifice. Again, we are faced with a piecemeal approach to the problem.

In addition to more technology and more money, we often hear that what we need is more and better laws which will set stringent standards to control the polluters. The banning of DDT in several countries and several states in the United States is no doubt a laudable victory in the war to save the environment. But it would be disastrous to mistake a single victory for winning the war itself. DDT is only one among several pesticides and herbicides which are detrimental to the ecosystem. Broader, more comprehensive laws are required—but they are slow to evolve and difficult to enforce.

The environmental "teach-ins" on college and high school campuses indicate that the young are aware and concerned about the problems. Many are convinced that an informed and educated citizenry will voluntarily change its traditional style of life and structure a society which is ecologically conscious. Others believe that the present styles of living are essentially unchangeable. Still others believe there must be a concerted pressure for change by activistic environmentalist groups. Luther Gerlach of the University of Minnesota describes a "movement" in ecology which is being fostered by activist groups throughout the country. He states that "the politics of environment will be the biggest mass movement in the history of the country." [12] But already there are signs of a potential backlash. Some fear that too much concern over environment will divert attention from more familiar problems such as poverty and race. Others feel that business and industry view environment as another campus fad which will soon pass. Still others are fearful that the movement in environment is subversive and has strong revolutionary overtones.

The efforts being made presently are enormous and will continue perhaps indefinitely. But the dimensions of the problem have not as yet been fully appraised, and the likelihood that many of these problems will increase is almost certain. More technology, additional funds, better education, new laws, and greater citizen involvement are needed to meet the crisis. But beyond this we must develop a new environmental ethic that emphasizes the symbiotic relationship between man and the Earth. Perhaps a new theology is required which gives value to nature and places man and spaceship Earth within the context of the cosmic universe. But a new ethic and a new theology must retain ecology's holistic view while at the same time they must help set social priorities which are life-supporting and life-affirming.

Ecology and Survival

It has become a common exercise for many individuals and a few institutions to spend a great deal of time thinking about the future. This new interest in the things that lie ahead is not unique to our century, but the number of publications, organizations, and conferences dealing with the near and distant future suggest that it is becoming a major preoccupation of our time. Perhaps the growing number of "futurists" share the opinion of Dutch sociologist, Frederick L. Polak, "Thinking about the future is not only the mightiest lever of progress but also the condition of survival." [13]

A great deal of work related to future trends is now in progress. There is an attempt to bring about a broad consensus as to what our goals should be as a nation.[14] And there is general agreement that whether we plan or don't plan, we are "shaping" the future for good or for ill. Much of the present planning is being done to "accommodate" the growing population, the growing urban concentration, and the growing demand for goods and services for all citizens. Growth and technological change are seen as inevitable and constant.

There is a myth surrounding the future. We continue to operate as though we can go on with business as usual. We assume that we can meet the increasing demands of a growing population by continually increasing the production and consumption of goods. But ecologists say, "No!" There are definite limits to growth of all kinds and we will soon reach these limits in many parts of the Earth. Yet we continue to operate as though the Earth's environment were infinite. Our hearts are set on "new frontiers" we long to explore and exploit. When there is no more land to conquer, our ambitions will be directed toward the oceans and outer space which we have already begun to consider.

Ecology is a relatively new word that has come to us from the biological sciences. The word first appeared in English in 1873. It is derived from the Greek word, *oikos,* which means house or habitat. *Economics* comes from the same Greek root, as does the word *ecumenical.* A great deal of knowledge is available on the interrelationships of rodents, insects, and birds to their respective environment. Ecologists can usually predict what will happen if the food chain is broken in any given ecosystem. But until recent years the major effort in ecology excluded the human species from its study. It operated as though men were not a part of the natural system; as if men stood totally apart from the balance of nature and the web of life.

Most ecologists today would view the population explosion and the famine in India and Africa as ecological problems. There are too many people living on too small a land area. Some ecologists believe that the land mass of the United States can adequately support a population of about 150 million. We now have over 200 million and the prediction is for an additional 100 million by the year 2000 or shortly thereafter. Those who object to the claims of overpopulation in this country argue that certain vast areas of the nation are virtually empty of human habitation and could be used for settlement purposes. But their views reflect a popular naivete over the ratio of living organisms to a specific land mass. Whether an area is settled or not, it is contributing to the support of the population in general. A large population needs space for growing food, open space for recreation, and space for the renewal of resources. We often forget that our somewhat more than 200 million citizens could not continue at the present living standard without using large amounts of resources from other countries. If we were forced to exist on what can be produced from within our own borders, we would think more seriously about what in the United States is an adequate and optimum population size.

An ecologist would also view urbanization as a major human ecological problem. The high concentrations of people in urban areas is viewed with alarm because of the resultant air pollution, congestion, lack of open space, noise, crowding, rising crime, and urban sprawl. It is difficult to imagine that at the present time 130 million people are living in large cities whose total land area covers hardly more than one percent of the United States.[15] Some predict that in the future 98 percent of the population of this country will be living on only two percent of the land.

Kingsley Davis, one of the leading authorities on urban development, has recently described the situation of the world's cities. He indicates that 38 percent of the world's population is already living in what are defined as urban places. Over one-fifth of the world's population is living in cities of one million and over. On present trends, it will take only 15 years for half the world's population to be living in cities, and in 55 years, everyone will be urbanized.

Davis foresees that within the lifetime of a child born today, on present rates of population increase, there will be 15 billion people to be fed and housed—over four times as many as there are now. The whole human species will be living in cities of one million and over, and the largest city will have 1,300,000,000 inhabitants. Greek planner, Constantinos Doxiades, calls this situation *dystopia*—a disease-ridden condition in direct opposition to utopia. "We do not need a crystal ball to foresee what Davis and Doxiades are predicting," writes Ritchie Calder, "we can see it through smog-colored spectacles; we can smell it 70 years away because it is in our nostrils today, a blind man can see what is coming."[16]

At the present United States birth rate, during the next 30 year period, our country's population will increase by 85-100 million. If we were to accommodate the full 100 million persons in new communities, it has been estimated that we would have to build a new city of 270,000 persons each month from now until the end of the century. This would mean constructing a city the size of Tulsa every 30 days for over 30 years. The impact of such rapid growth and the accompanying demands on all resources and all facilities is impossible to imagine. We can assume that if such growth does take place, the total environment of this nation will be in serious jeopardy. The amenities which we now enjoy will indeed be sparse at the turn of the century. Any person concerned about the quality of human life must view the prospect of an additional 100 million with great dismay.

Whatever is done to the environment is likely to have repercussions in other places and at other times. The environment is finite. Our resources are finite. When the supply runs out we will have to recycle what we have used. The capacity of the environment to serve as a "sink" for our total waste, to absorb and recycle so that it does not accumulate as pollution, is limited. In many instances that limit has already been passed. It seems clear that when limits are passed, gross effects occur, some of which are predictable, but some of which are not.

Ecology must convince the world's population that the only solution to the problem of growth is to stop growing. This applies to population and, unless the population is on the decrease, to its standard of living. It should be quite clear by now that "standard of living" is beginning to have an inverse relationship to the "quality of life." A constant increase of gross national product must be viewed from an ecological standpoint as disastrous. This does not mean that the ecological point of view is against progress. It calls into question whether progress can be defined narrowly as the *quantity* of possessions available to a population. Ecology is ready to sacrifice quantity of goods for survival of the species, all species, and it would opt for the essential *quality* of the physical and social environment.

Ecology places harsh demands on mankind. Its laws tend to subvert the "growth syndrome" of many nations and cultures. It has just begun to criticize the careless activities of 20th century civilization. The worldwide saturation of DDT and other pesticides has just recently come under the careful scrutiny of the ecologist. We are being told by many eminent scientific sources that our social values and life styles must be radically changed. We are being confronted by an ecological ultimatum: either we redirect our priorities as a world society, or run the risk of extinction as a species, or reduce the physical and social environment to an artificial and sterile habitat, lacking clean water, pure air, natural beauty, or places of tranquility.

There appears to be an implied assumption that if the Earth becomes too polluted or too overcrowded, we can simply move on to other planets. But anyone who has studied the dimensions of space in relation to the distances that modern day astronauts have traveled will know that the moon can hardly be considered a beginning. The cost and preparation of sending large numbers of humans to inhabit other unknown planets may be possible in the far distant future, but it cannot happen unless we are able to survive as a species through the next three decades and well into the 21st century. Some doubt seriously whether man will ever actually travel in space and live on other planets. René Dubos has stated, "Science-fiction writers and a few scientists notwithstanding, man will never be able actually to settle anywhere in the cosmos other than on or near the surface of the earth." [17]

The future has "limits" whether or not man ultimately travels extensively in space. This is the most important thing that ecology has to tell us. The environment is not infinite, and we must face the problems of the Earth's spheres, particularly the biosphere. We must think of the Earth as a closed system except for the intake of the sun's energy. The planet Earth has become one small world. Although man can reach out in space, he must realize that his life span must be lived out in this planet's biosphere for some time to come and perhaps indefinitely. If the space program has created for man a new image of himself and his Earth so that now and in the future he will devote himself to caring for the various Earth spheres, we can conclude that the investment in energy and money is truly minute. On the other hand, if the new image created is the myth that we can now treat the Earth as dispensable since we can now "escape" to other worlds, the space program will have been the most costly and shortsighted development in the history of man.

It is obvious that if man continues to persist on the planet, assuming that nuclear war, famine, or the plague do not ensue, he will be required to limit his numbers radically; to limit the exploitation of natural resources; to limit the growth of urban centers; and to limit or control the uses of technology. It will require a self-discipline not previously required. It will mean that individuals and nations must set strict boundaries on production and consumption and the disposal of waste. It will require some form of international law to monitor human reproduction and pollution. It will mean that man will be forced to move forward without physical frontiers. This, it seems clear, is a crisis never before experienced by mankind on a global scale. Learning to live with the finite, within boundaries and limitations of the Earth's resources and the Earth's systems, will be one of the most difficult tests for man.

We are rapidly approaching the "crisis of the finite" in this and other lands. The United Nations Conference on the Human Environment scheduled for Stockholm in 1972 reveals that the situation has become critical for many of the world's governments. The solution to environmental problems cannot come unless the nations of the world are willing to exercise the basic principles of ecology on a "macro-ecologic scale." Competition must give way to cooperation so the great organism, Earth, can continue to harbor life. We in the United States should initiate the reassignment of priorities, especially since we appear to be the planet's chief "parasite" among the nations of the Earth. But we will not be successful unless we can mature sufficiently to live within the confines of a finite world. The crisis of the finite will be a testing of man's essential nature, but it will mainly be a testing of the American pioneer spirit.

Although it may not be obvious that Americans operate on a frontier mythology, it is becoming increasingly clear that we are moving toward a point in human history when we must limit most aspects of growth and production. We cannot indefinitely produce automobiles and people at the present rate and expect not to be smothered to death by either or both. And we cannot continue to exploit the resources of the Earth without thinking about the needs of future generations. We must learn to think of ourselves and our institutions as part of a massive global system of life. And we must learn to deal with our problems in a comprehensive way. In order to do this, we must gain a new perspective on the "total" environment of man, not simply of litter along the highway or the pollution of one lake. We must see all problems in the context of the whole Earth and become aware of what man is doing to threaten the future of our planet. James Russell Wiggins dramatized our precarious situation when he stated:

> *We are in a sense a quarreling crew on board a sinking ship, disputing over the breakfast menu while the water in the bilge rapidly rises and the great vessel sinks slowly into the vasty deep.*[18]

A VIEW FROM SPACE

2

The universe may
be as great as they say.
But it wouldn't be missed
if it didn't exist.

—Piet Hein

If the titles *Silent Spring*[1] and *The Quest Crisis*[2] suggest the vociferance of the present revolution in man's environment, we might conclude that the ecologic crisis is the most quiet of the quiet revolutions. Wars, riots, and killings daily appear on the front pages of most newspapers, while radical changes taking place in the environmental envelope surrounding the Earth go virtually unnoticed except for an occasional editorial on the evils of air pollution, or a letter from an irate citizen suffering insomnia from low-flying aircraft.

We are far too preoccupied with the "predatory uprisings" within the human sector to be interested in the rapidly decreasing numbers of certain wildlife species or the encroachment of DDT on the world-wide food chain. Nor do we become alarmed over the mounting quantities of waste products which pollute our rivers and streams, or the growing menace of carbon monoxide spewed from automobiles which each of us drive. Perhaps it is the "illusion of central position" described by Robert Ardrey in his book, *African Genesis*.[3] We have epitomized the human sector and the human drama at the exclusion of other forms of life. But even more serious is the total lack of perspective we exhibit as we view ourselves in relation to the Earth, and the Earth in relation to other celestial bodies.

Our main problem may be one of shortsightedness—a literal shortsightedness. We are standing too near our object, and we spend most of our time contemplating the human navel. We do not become sufficiently concerned because we are looking at individual trees rather than the entire forest. We become alarmed only when the lake where *our* cottage is located becomes seriously polluted, or when the thundering jets begin to fly over *our* home.

We urgently need to deal with large segments of the forest and to view changes that are taking place from a broader perspective. It is necessary, then, to take several giant steps backward to gain a holistic view of the environmental crisis and to understand individual problems in relation to the total environment of the Earth. Then, and only then, are we able to see the full implications of the environmental revolution which is taking place.

Dimensions of the Earth

The future astronaut can be compared to a small boy paddling on the shores of a large ocean. When he has reached the moon, the boy will have waded out to the nearest rock. When he has attained the farthest planet of the solar system, the boy will still be no further than the offshore islands within view of the beach. Only when the astronaut goes beyond the solar system and enters the next star system nearest our sun, will the small boy have reached a dangerous distance from shore. The Earth is tiny compared to the solar system. But the solar system compared to the immensity of the universe is infinitesimal. It is large only in terms of Earth-measured miles and Earth-determined time.

Ritchie Calder points out that astronauts circling the globe 16 times a day managed to do what generations of cosmologists had failed to do—to convince the ordinary layman that the Earth is a small planet in a vast cosmic system.[4] We have been forced to modify our traditional concepts of space, time, gravity, and acceleration. When news stories described the first launchings of space satellites, we began to appreciate the meaning of momentum. The buildup of speed to 17,500 miles per hour was necessary to offset the gravitational restraints of the Earth in order to put a capsule into orbit. This taught us that we inhabit a tiny spacecraft which is traveling at an amazing speed through space.

The Earth is one of nine planets in the solar system. It is nearly 8,000 miles in diameter and the highest peaks are only about 12 miles above the ocean floor. The Earth is nearly 25,000 miles in circumference and it rotates at approximately 1,000 miles per hour on its own axis. Water covers more than 75 percent of its surace, and only a portion of the remaining land is fit for human settlement.[5]

We live on a small planet bound to a medium-sized star. The sun is on the outer edge of one of billions of galaxies which make up the universe. Viewed in this way, from the depths of outer space, our Earth is but a tiny speck. It would be easy to miss if it were not for the envelope of life that grows on its surface. Of all the places in the universe that conceivably could support life, the Earth is the one and only place we know of that does. Until we can communicate with planets similar to our own in other star systems, we can only assume that the Earth is the life-center of the universe.

French paleontologist, Pierre Teilhard de Chardin once stated, "The world is round so that friendship may circle it."[6] The very fact that the Earth is round is important; the "sphericity" of man's environment is significant. In an unlimited, flat environment man's thoughts and social interaction would simply diffuse outwards and be lost. But when it is confined to spreading out over the surface of a sphere, idea will encounter idea; the result will be an organized web of human thought.

Whether we are aware of it or not, we are all riding on a little spaceship which is so uniquely designed that, so far as we know, human beings have been living on it for over two million years not even knowing they were on board a ship. It is this tiny sphere which we are so busily exploiting: using up its oil and coal deposits; polluting its lakes and oceans; spreading urban development over its surface; chopping down its forests; and slowly filling its air envelope with various poisons. It is the only "ship" in the solar system which can support human life. Other life-supporting planets in more distant star systems would be impossible to reach. We have only one choice: to preserve this Earth so that life can go on.

Journey Around the Sun

As space distances go, the moon is nearly at our back door—a mere 239,000 miles away. In terms of distance from the Earth, Mars is to the moon what the Empire State Building is to a pebble. At the nearest approach to Earth, it is 35,500,000 miles away, which represents nearly a two-year journey for a present-day astronaut. The sun, nearly three times that distance from the Earth, would take more than a five-year journey one way. If the Earth were represented by a point one foot from a second point representing the sun, Pluto, the most distant planet in the sun's system would be no less than 40 feet away. The solar system including the sun and its nine planets is seven billion miles in diameter.

The Earth, spinning on its own axis at 1,050 miles per hour at the equator, is also revolving around the sun once each year in a 600-million-mile sweep that takes almost exactly 365¼ days and is responsible for the succession of the seasons. The Earth moves about the sun at 18½ miles per second, 1,100 miles per minute, and 66,000 miles per hour.[7] The orbit around the sun is not a perfect circle, but an elipse; the Earth's average 92.9 million mile distance from the sun varies by 3,100,000 miles during the course of a year. The variation between winter and summer, however, has nothing to do with this alternate nearness to and distance from the sun.

The sun, containing nearly 99.9 percent of the substance of the solar system is the nucleus of the system and controls the movements of the planets and other bodies. Not only is the sun, with a diameter of 864,000 miles, much larger than any planet, but it is all gas—all two billion, billion, billion tons of it.[8] No other object in the universe is as important to man and the Earth as the sun. It is the central fire on which depends all life on the Earth and any life that may exist elsewhere in the solar system. Its surface consists of a seething mass of hot gases and subatomic particles with an average temperature of 10,000 degrees fahrenheit.

Our sun moves in company with the Earth at just the right distance to provide us enough radiation to keep us alive, but not close enough to burn us up. And the whole scheme of our spaceship is so well designed that magnetic belts (Van Allen) filter radiation impinging upon our spherical home in such concentration that it would kill anyone outside the belts. This vast, doughnut-shaped band of radiation that girdles the Earth is up to 40,000 miles thick and is a consequence of the Earth's magnetic field.[9]

We now have some perspective of the velocities and movements through which we pass on our "journey around the sun." We are not conscious of these movements in our daily routine of life, but we are totally dependent on the proper functioning of the system. The potential hazards caused by global environmental changes provoke fear in all of us. Seeing the Earth in the context of the solar system reminds us again of the uniqueness of our home and the importance of maintaining it in good order.

100 Billion Stars Make a Galaxy

We are conditioned to think of ourselves as the center of the universe. We believe that the universe including the stars, sun, and planets rotates around our home, the Earth. We also operate under the illusion that our Earth is massive. A trip around the world even today seems an immense journey. But placing the Earth and its movements into proper context within the solar system reveals that we are moving at extremely high speeds in an unbelievably broad sweep in nearly empty space. In one year our Earth covers billions of miles in its trip around the sun. To have a sense of our pear-shaped home carrying us around in space is a major step in helping to become oriented in time and space. But to assume that we now have described our place in the cosmic scheme of things would be a gross error. There are other movements and other distances that must be incorporated into a proper view of ourselves and our spacecraft.

The solar system, the sun and its planets, must be seen in another context. The sun is one among an estimated 100 billion stars that make up the Milky Way galaxy.[10] The distance within our own galaxy is so vast that the entire solar system, including planet Earth, is reduced to microscopic scale. If we applied the scale (Earth and sun set one foot apart, or one foot equals 93 million miles) to the distances outside our own solar system, the nearest star would be at a point 50 miles away. The nearest star to the sun, Alpha Centauri, is 24 million, million miles away. In terms of light years (the distance that light travels in one year; light travels at 186,234 miles per second), Alpha Centauri is 4.3 light years distant. When we see this star on a summer night, we are actually seeing light which is 4.3 years old.[11]

Our sun and Alpha Centauri are parts of a disc-shaped cluster of stars estimated to contain 100 billion stars, some larger and some smaller than our own sun. This disc-shaped cluster is the Milky Way galaxy most of us can observe during summer nights directly overhead. The Milky Way is so vast that we have difficulty in comprehending its immense size. Light traveling at the velocity of 186,234 miles per second would take 100,000 years to cross its breadth. It is 100,000 light years in diameter and 25,000 light years thick at its center.

The sun with its planets is located at approximately two-thirds the distance from the center of the galaxy. It is moving at 12 miles per second in the direction of the constellation Hercules, and at 150 miles per second around the center of the Milky Way galaxy; a circuit that takes 200 million years to complete. Since the beginning of man, our system has moved less than one percent of the total revolution around its circuit.

If we compare the space occupied by the planets with that occupied by all the stars of our galactic system, it would be like comparing one square foot with the whole area of North America. The Milky Way is a disc of stars spanning a diameter of an inconceivable 600 million, billion miles. Compared to such distances, the Earth's diameter is truly small. Those vitally concerned for the environment need to view the Earth in terms of these vast cosmic systems. We must operate with a new perspective of the whole Earth, rather than a piecemeal approach to individual parts.

100 Billion Galaxies Make a Universe

If we could conclude that the Milky Way is the extent of our universe, we might learn in time to become comfortable with its size and shape. But our galactic system is only one of what has been estimated to be 100 billion galaxies making up our universe.

It appears that much of what we observe is moving away from us at a very high velocity. The constellation Virgo, composed of a cluster of several thousand galaxies, is about 50 million light-years distant and is moving away from the Milky Way at 750 miles each second. The Corona Borealis at a distance of about 940 million light-years is receding at a velocity of 13,400 miles per second. A cluster in the constellation Bootes at a distance of 1.7 billion light-years is moving away from us at 24,400 miles per second. And in Hydra a group of faint galaxies at a distance of 2.7 billion light-years is receding at 38,000 miles per second. Other more distant clusters are thought to be moving at nearly nine-tenths the speed of light. If any galaxies move outward at the speed of light, it is doubtful that we will ever be able to observe them.[12]

The basic unit of the universe is the galaxy, like our own Milky Way. Billions of galaxies are racing through space, out and away from one another toward an unknown destiny. Only two galaxies are visible to the naked eye. The Andromeda galaxy sends light to the eye from a quintrillion mile distance. It is trillions of times more distant than the moon, billions of times more distant than Pluto, over two million times more distant than the nearest star, Alpha Centauri, and over a thousand times more distant than the faintest star visible to the naked eye. When we see the Andromeda galaxy, we are seeing light more than two million years old.

Life on a Tiny, Whirling, Celestial Orb

An imaginary journey beyond our own atmosphere, beyond the solar system, and beyond the borders of the Milky Way into the outer reaches of the massive universe, leaves us with a sense of insignificance as we see ourselves inhabitants of a tiny, tiny globe. Our Earth's surface contains approximately 36 billion acres. It is difficult to acquire a personal attraction toward something so vast. "Yet when compared to a billion-light-year universe," writes S.P.R. Charter, "our Earth is smaller than a hummingbird's wingtip against the mountain barrage of all the Rockies." [13]

With his vast knowledge of physics and biology, R. Buckminster Fuller has come to think of the Earth as man's spaceship. His *Operating Manual for Spaceship Earth* [14] indicates the degree to which he conceives of the Earth as a small "ship" and mankind as its "pilot." Infinitesimal in relation to other masses and spatial distances, this small self-contained sphere is darting through space toward an unknown destiny, except for its annual orbit around the sun and the larger circuit around the hub of the Milky Way. One slight inclination in its pathway around the sun could turn man's spacecraft into a fiery furnace or a sub-zero freezer. But the system seems to work century after century, and the thin veneer of life which encircles the globe grows ever more mystifying.

We must develop a "feeling" for the entire Earth—to think of it as our own spaceship—a nest in which the human species was born. We are now at a turning point in the evolutionary process. Natural selection is giving way to human manipulation. The future is grounded more and more in man's decisions, and his concept of who he is and what his task is becomes ever more important. Is it possible that the present space explorations are the beginning of man's first baby-step from his nest? Is the Earth simply an incubator where the life envelope is giving birth to an "intelligence" which will serve as a "pilot" for the whole known universe? This will not take place in the near future, but what of the long-range future, another million years?

It is important that as many of us as possible gain a new vision of the whole Earth. Perhaps it is the vision of Adlai Stevenson who once stated:

> *We all travel together, passengers on a little space ship, dependent on its vulnerable supplies of air and soil; all committed for our safety to its security and peace, preserved from annihilation only by the care, the work, and I will say the love we give our fragile craft.*[15]

Or it may be the vision of Archibald MacLeish who wrote at Christmastime in 1968:

> *To see the earth as it truly is, small and blue and beautiful in that eternal silence where it floats, is to see ourselves as riders together, brothers on that bright loveliness in the eternal cold—brothers who know now they are truly brothers.*[16]

PERSPECTIVES FROM ANTIQUITY

3

What we have to face is that the insect is a good three hundred million years older than we are.

—C. K. Brain

The view of Earth from space will convince us unequivocally that our planet is indeed infinitesimal. Its dimensions and surface area are no match for the incomprehensible distances in the universe. The size and structure of our globe in the context of the cosmos should quickly impress upon us a new appreciation for its delicacy and frailty. The present disturbances which are occurring on its surface as a result of man's activities must be evaluated in terms of the Earth's uniqueness as a planet. We do not know whether there are other "earths" in the universe. In terms of mathematical probability, there are thousands of chances that planets similar to our own do exist.[1] But we cannot be certain, and until we are, we must give our Earth the prominence it deserves.

Another view is equally as important as the view from space. This is the view from the distant past. It involves a backward look—back to the origin of the planet and to the origin of life on the planet. We are in the habit of being grossly shortsighted when it comes to looking backward. We commonly measure the past in terms of human history excluding the eons of time which preceded our own emergence on the surface of the globe. A perspective from the distant past can help us better appreciate the value of our Earth and illumine the crisis facing the environment.

Beginnings of the Earth

It is unlikely that Archbishop James Ussher would today date the moment of creation as precisely or as recently as he did in 1650. Then he dogmatically identified the event as taking place on Sunday, October 23, in 4004 B.C.[2] So acceptable was his candid conclusion that at one time this date was placed in the marginal notes of the King James Version of the Bible. There it stood for nearly 300 years as an indisputable bench mark of fundamentalist theology.

For centuries the Bible has been thought of as witnessing to a small geocentric and static world governed by a wise and almighty God whose main interest is to help man achieve his eternal destiny. Now, however, man looks insignificant against the background of the vast dimensions of time and space. Man is said to have emerged out of natural evolution governed by a blind and arbitrary power. Has the God of the Bible any relevance for this new world view? Can we still speak of the intimate relationship between the Creator and his creation?

The Hebrew spoke affirmatively that God is the Creator of the universe. He is the sole source of everything that exists. In fact, there is no material substance which can exist independently of the creative will of God. He was in existence prior to everything which now exists.

The doctrine of creation asserts that God is the Creator of all things, including all processes that have brought about the evolution of man. His creation consists not only of man and the Earth, but includes the planets, stars, and galaxies. The entire universe was made by him.

For modern man nature is no longer a static, geocentric, limited entity, but a process in an almost infinite space and an almost endless time. The Earth is a tiny satellite of a small star in one of the many billion galaxies discovered by modern telescopes. The evolution of nature went through a long series of changes and events: first as elementary nature, then as simple life, and on toward greater and greater complexity until conscious life appeared in the form of man.

Various theories have been advanced to explain the creation of Earth. Most agree that a probable chain of events occurred, either over long periods of time or in a single explosive event that resulted in the spherical ball we now call planet Earth. Religious sources of the Judeo-Christian tradition disclose that all occurred as a result of a divine plan and intent. But in no instance do these sources claim to know the exact process used or the sequence of events.

While knowledge of the Earth's size and shape is ancient, man's understanding of the planet's origin is yet very imprecise. The beginning of the Earth dates back billions of years. Most geologists today accept the figure of 4.5 billion years as the present age of the planet. But there was a time when the Earth did not exist. And the entire question of its beginning is related to the larger question of the origin of the entire solar system: of the stars, galaxies, and of the universe as a whole.

We can easily state the approximate age of the Earth, but we cannot begin to comprehend the enormity of time that has transpired since the Earth was born. A thousand years is nearly impossible to imagine, and a million years is far beyond our ability. The Earth is nearly "five thousand times one million" years old.

The present revolution taking place in the environment must be viewed in relation to an ancient Earth. We are not dealing today with natural systems invented during the Industrial Revolution or conceived by 20th century technologists. The Earth presents us with a living laboratory of the far distant past. Much of what constitutes the universe is represented in our Earth. We often think that the universe is somehow "out there." We must remind ourselves that the planet on which we ride is an integral part of the universe and may represent an advanced stage of cosmic evolution.

Emergence of Life

The beginnings of the Earth and the emergence of life on its surface represent two significant "points" in our geologic history and perhaps in the history of the universe. The Earth made hundreds of millions of journeys around the sun before there was any evidence of elementary life. Not only is the beginning of the Earth hidden in the far distant past, but the origin of life and the exact conditions under which it first evolved are yet only vaguely apparent to us.

When Charles Darwin began his famous five-year world cruise aboard the *H.M.S. Beagle,* he was well aware of the prevailing theory of the origin of life.[3] Having failed in his study of medicine, he was sent to Cambridge University to prepare for the ministry in the Church of England. The popular view of the origin of life had become a firm and unquestioned doctrine. Every species of plant and animal had come into existence at one grand moment; even the hour of the day had been prescribed.

Darwin did not dispute this theory. Learning that he would eventually inherit sufficient property to insure his future, he devoted himself to observing and wondering about relationships of living things and the origin of life. As an unpaid naturalist aboard the *Beagle* on her second major scientific expedition, Darwin was well prepared for the discovery of new knowledge. His greatest discovery was evolution, the concept that there is an essential "kinship" among all living things because all have emerged from one common ancestry. He challenged the old view of creation with his two books, *On the Origin of Species* published in 1859 and *The Descent of Man* published in 1871.[4]

Darwin tried to avoid the prejudices against his own findings by not discussing the origin of man, but he also did not want to deceive a few unprejudiced persons by concealing his views. He settled the problem by adding one meaningful sentence to his concluding chapter: "Much light will be thrown on the origin of man and his history." It was this one sentence that aroused the skeptics and drew condemnation on himself and his work from many authorities; among his harshest critics were the clergy.

The essence of Darwin's work sets forth the principle that there is a struggle for existence, and in this struggle the "fittest" survive. But the struggle to survive does not imply an all-out war of life against life. There exists a dependency of one life form on another, and not only is there an instinct to preserve the life of the individual, there is also a concomitant drive for success in leaving offspring to continue the species. A cooperative as well as a competitive process exists in which life struggles to perpetuate itself. Wherever one finds life it will be accompanied by an internal force to move forward beyond itself. Although one may find an individual animal or a species becoming extinct, he will also find elsewhere a new surge—life again promising to spring forth in new and more creative ways.

Darwin had explained not the origin of life, but the "process" of life. All of life is one because all life has arisen from one solitary, "unremembered beginning." The process does not represent a simple, clear-cut progression easily traceable and understood. Rather than a deliberate climb of a "flight of stairs," the process better resembles the randomness of the "growth of a tree," branching out in this and that direction leaving only bits and pieces buried in the fossil remains. The climb of life to its present state is better represented through the abstract brush strokes of an artist than through the deliberate tracings of a draftsman. The "theory" of evolution cannot be refuted on grounds that we now fully understand how life emerged on the Earth. The attitude that evolution is simply a theory is readily negated by the indisputable remains of plants and animals which lived and died during this long climb from the first organized form to the immediate predecessors of man.

Although our planet and the rest of the solar system is estimated to be nearly five billion years old, for much of this enormity of time the Earth was lifeless. The very oldest rocks have been dated back as far as 3.6 billion years and have no trace of fossils.[5] Life began possibly more than two billion years ago. How it began and where it began is not fully known. The account in the first chapters of Genesis is a statement of faith that God is the Creator of everything in the universe including man. It is not a factual statement of how and where life, including human life, began. Yet somehow, life did emerge on the Earth. Perhaps it is the only life in the universe. It may be the only life process that has produced a "man." Some speculate that life on Earth represents a trend toward which the entire universe is moving.

The idea that man and the world he inhabits are products of an evolutionary development seems degrading to some people. It is not degrading, but adds to the dignity of man. The fact that humans have grown out of the natural systems speaks to the marvel of the creative process and places new value on the diverse forms of life that have made humans possible. Knowledge of evolution provides us a new perspective on the unity of life and the interdependency required to maintain the life systems.

The process of evolution informs us that the world is not static, not fixed, not finished, and not unchangeable. Everything is engaged in evolutionary flow and development. Theodosius Dobzhansky writes, "Evolution . . . is a source of novelty, of forms of being which did not occur at all in the ancestral states."[6] This view is consistent with the biblical view of creation. Creation is not a single incident that can be mathematically dated. Rather, it is the ongoing activity of the Creator who participates in his creation. If such activity did not take place, there would be no creation. The Creator and his creation are inextricably bound together.

The obvious uniqueness of the Earth in our corner of the universe is of critical importance as we observe the present revolution in the human environment. We must carefully examine the present effort to build a high level of technology in relation to preserving irreplaceable aspects of the natural environment. We must ask whether gross national product will continue to be the prime indicator of quality of life, or whether there is another, higher principle that encompasses economics and technology. The "natural" riches of the Earth, including man, are the result of a long, unique process which for us seems to represent no specific design or plan. The mysteries surrounding life's origin may yet reveal themselves as the process of evolution continues. It appears that the new responsibility for man in this century is to insure the continuation of evolution.

Man as an Infant on the Earth

It has been suggested that "if all of geologic time were equal to one calendar year, the presence of man on the earth would account for one-half hour, and man's recorded history for one and one-half minutes."[7] We are in the habit of thinking of man as ancient, reaching back to the very beginning of the Earth's formation. He is, in fact, an infant on the Earth. His two million year history is but a speck of time compared with the nearly five billion year old planet. In terms of the human species, life still has before it long periods of geologic time in which to develop. Life in its "thinking form" still shows every sign of growth and expansion. Compared with former layers of biological life, mankind is so young that it could be called "new-born."

Man is the most recent to become dominant among land animals. He appeared in the envelope of living matter equipped with all the biological advantages characteristic of mammals. But superimposed on those advantages was the power of conceptual thought; a distinction that no other animal possessed before him. But before such a creature could be thinkable, a multitude of interacting forces, including living things, had to build up a supporting matrix or structure. The hierarchy of life had to start from the beginning with the simplest microscopic cells which step by step established the higher forms. All around us we see a world which has in the course of millions of years developed from inert materials.

Teilhard de Chardin characterized the whole of his life's work as an attempt to "view man within the evolutionary perspective wherein science has placed him." His primary interest was to discover whether man's development on the face of the Earth was unique. Did it have nothing at all to do with the natural evolution, or did the emergence of man represent in some sense the ways of the universe, the major trends, the meaning of the universe, or at least a part of the meaning? Teilhard wrote: ". . . however solitary his advent, man emerged from a general groping of the world. He was born a direct lineal descendant from a total effort of life. . . ."[8]

Teilhard's world of discourse was the breathtakingly massive universe and the appearance within it of the phenomenon called "man," that point in the evolutionary process where the cosmos begins to "think and steer itself." He believed that man is not the center of the universe as once we believed in our simplicity, but something much more wonderful—"the arrow pointing the way to the final unification of the world in terms of life." Teilhard was convinced that the advent of man represents the "last-born, the freshest, the most complicated, the most subtle" of all the successive layers of life.[9]

The awesome time span of evolution before the "flowering of life" and the appearance of a reflective being is an indictment against the 20th century's worship of immediacy, routinization, and technological change. Modern man has forgotten how he came to be what he now is. As a result, he cannot look hopefully toward the future. He does not see evolution itself as a source of hope.

When we have caught a vision of the Earth's antiquity —how long the journey has been for man—how complex the process that has produced all of us—we can begin to appreciate the potentials for further evolution.

What lies ahead? How will mankind develop from his infancy into adulthood? How might we individually and collectively insure that the life process will continue beyond this century? How can we avoid the dangers inherent in the ecological crisis? Overspecialization? Extinction? Entropy?

A Proper Sense of History

Teilhard de Chardin testifies that ". . . it is not only man but the world itself which is *history.*" [10] We are in the habit of equating history with human history, and more precisely, with the history of civilization or "recorded" history. This limited view distorts the reality surrounding our ancient world and leads to the conclusion that man's origin and the origin of the Earth are coincidental. A proper sense of history should encompass the entire span of time representing the age of our Earth. We should gain a new appreciation of the patience with which life emerged on the planet, and we should become acutely aware of the extremely brief period which represents "human" history—and the speck of time which is included in the term "recorded" history.

We are now the dominant species among plants and animals. Our species is yet extremely young. The possibilities for the future are enormous. The greater part of our life as a species lies before us. We are essentially at the beginning point—only at the threshold of the potentials for man. And yet, our very existence is threatened from overpopulation, nuclear war, overspecialization, and ecological disaster. Can we afford to continue to perpetuate our increasing exploitation of the Earth's resources if it means that future generations will experience a serious decadence? Can we afford to change our patterns of growing population and growing affluence at the risk of depleting the growth of humans as well as other forms of life? What are we willing to sacrifice in favor of the future? Do we want history and evolution to continue? Could the future now rest in our hands and be based on our decisions?

THE ANIMATE EARTH

4

. . . respect for the kind of intelligence that enables grass seed to grow grass; the cherry-stone to make cherries.

—Teo Lei
15th Century

Astronomers describe the Earth as just another planet with no special distinction among all the countless celestial bodies. The Earth's uniqueness is tied to the fact that she transcended her material nature when she began to harbor life more than two billion years ago. Without life, the Earth would be just another drab fragment in the solar system. Fortunately, however, living things have converted the inanimate matter of her crust into a living substance. The infinite beauty of her blue atmosphere and green surface are a result of the activities of the countless microbes, plants, and animals that she has nurtured and which have transformed her substance. We humans exist and enjoy the world only by virtue of the environmental conditions created and maintained by all other forms of life. The Earth is a home suitable for man because her crust is literally a living organism.

The Earth is not dead as we often assume. It is alive—a living, pulsating, cosmic body. The Earth never remains constant. It is forever changing. The process of creation continues to produce fresh and ever novel conditions. The most recent innovation is man. From the perspectives of cosmic space and geologic time, the great events of today are the growth and activity of man on the Earth's surface. And he, too, is undergoing radical changes which can be perceived only in retrospect thousands of years hence.

That the Earth is alive is of particular importance to our period in history. Our technological prowess creates the myth that only man is alive, while all non-human substance is considered dead or valueless, until measured in terms of man's need. The natural environment is viewed essentially as a resource—for man. It is, in fact, a living, pulsating body, displaying ever greater complexity and originality. The present saturation of the environment with numerous pollutants is a serious insult to the living planet. The ultimate cost of our unswerving urge to produce, to consume, and to grow will be borne by the Earth itself, and indirectly by the life she carries.

Spheres of the Earth

If the ecologic crisis is a serious threat to the survival of the planet, perhaps it was worth the several billion dollars spent on the Apollo missions to make us perceive through the eyes of the astronauts the color and inviting beauty of the Earth in contrast to the bleakness of space. Now that we have experienced the sterility of the moon's environment, we can express a greater appreciation for our kindly home. And now that the Mariner space probes have dispelled our illusions about the existence of life on Mars, we can place a greater value on the uniqueness of the Earth in the solar system. The view of the Earth from outer space reveals that we are living comfortably on a blue-green, sunlit, "oasis" which floats endlessly around the sun.

John Lear describes what is believed to be the "pulse-beat of planet Earth."[1] Although the beat cannot be measured, several manifestations of it are regularly observed. For example, a rhythmic throbbing in the magnetic shield enclosing the planet is caused by the solar wind and has been detected by spaceships sent from Earth. Within the magnetic covering, the Earth's atmosphere expands and contracts with regular heating and cooling of the sun as day passes into night and night into day. Below the atmosphere, the oceans rise and fall in fluctuations that must be estimated rather than measured because the crust on which we ride is also moving in response to the tidal pull of the moon. The implication of this rhythmic motion is to treat the Earth's oceans, as well as the atmosphere, as "one continuous, constantly interacting sea."

In a similar way, Teilhard de Chardin has described the Earth's crust as not yet having stopped ". . . heaving and plunging under our feet."[2] Even today large mountain ranges are being forced up on the horizon, and granites still grow under the continental masses. Lakes are dying and being born, rivers are slowly changing their course. According to Teilhard, the organic world also has not ceased to ". . . produce new buds at the tips of its countless branches."[3]

The pulsations of the Earth take place in a series of approximately concentric spheres including the magnetosphere, atmosphere, hydrosphere, lithosphere, and barysphere. Geologists describing the successive layers of the Earth speak of the barysphere, composed of metals; the lithosphere, composed of rocks; and the hydrosphere, composed of water. Biologists refer to the biosphere as that which contains all living things and includes parts of the atmosphere and hydrosphere.

Teilhard de Chardin invented a new term which he called the "noosphere"—taken from the Greek word "noos," referring to the mind.[4] Teilhard referred to this new layer or membrane on the Earth's surface as the "thinking" layer superimposed on the "living" layer of the biosphere and the "lifeless" layer of the inorganic material, the lithosphere. Kenneth Boulding has used "sociosphere" and "anthroposphere" to describe the same layer.[5] This is, of course, the sphere of man and all of his activities. In our time it increasingly dominates all the others and is the source of the ecologic crisis, as well as the potential force to restore balance and quality to the environment.

The various spheres which make up the system called Earth do not exist separately or independently from one another. In fact, they blend together as though they were all part of a total living organism. Even the noosphere or sociosphere representing man cannot be separated from the entire system of spheres which compose the Earth.

Nuclear explosives are believed to have some effect on the magnetosphere which extends to more than 40,000 miles from the center of the Earth. The pollution of the atmosphere, which extends to about 40 miles above the Earth's surface, will have some direct effect on the biosphere and certainly on the sociosphere from which the major portion of pollutants emanate.

The atmosphere is not only the air which humans breathe; it is the envelope which protects living things from harmful radiation from the sun and outer space. It is also the medium of climate, winds, and rain. These are inseparable from the hydrosphere, including the oceans with their currents and evaporation, and from the biosphere with vegetation and its transpiration and photosynthesis, and from the lithosphere with its minerals and deposits of ore.

It is a peculiar phenomenon of the present period to witness the growing threat of the sociosphere to all other spheres. Mankind as the fresh, new growth in the Earth's evolution, signals both a new direction and a potential end to this unique planet hidden in the depths of space.

Preeminence of the Biosphere

From the perspective of cosmic space and geologic time, the biosphere must be given the highest prominence. In the universe it is a delicate, almost imperceptible "skin" which has been growing ever more complex over the past two billion years. Man's advent and his power of reflective thought would indicate that the biosphere is also growing more "conscious" of itself.

All life is confined to the biosphere which extends down into the ground as far as the deepest growing trees, into the air as high as the tallest trees, and encompasses about 500 feet of the ocean's surface. Thin as the biosphere is compared to the overall dimensions of the Earth, this film contains more than one million kinds of plants and animals. Few of these life forms are spread throughout the globe. Most are limited to a specific area such as a continent. Only man has inhabited most of the land masses. But impressive as his accomplishments may be, man has never left his ecological home on a permanent basis. Whenever he does, he must carry with him water to drink, air to breathe, and suitable food for nourishment. For each space journey of the astronauts, it has been necessary to create the Earth environment within the sealed enclosure of the spacecraft.

The most radical changes which have occurred in the environment in the past few decades have taken place in the biosphere. Here it is not only the changes that have taken place within the human realm, but changes that have affected all forms of life. Numerous species of plants and animals have been or are being killed by pesticides or herbicides. Treasured birds, trees, and plants have virtually disappeared as a result of deliberate killing or radical changes in their habitat. The natural life-and-death cycle of lakes and rivers has been critically affected by man's pollution or manipulation. Tree-crowded landscapes have been bulldozed for housing projects and industrial development. Millions of acres of greenery are covered over each year for buildings, parking lots, streets, and freeways. Shallow lakes and swamps are filled in to make way for urban expansion in spite of their value as a wildlife habitat. Combustion of fossil fuels has produced a 10 percent increase in carbon dioxide in the atmosphere during the past century. More than one billion acres of arable land has been lost through erosion and salinization. Two-thirds of the world's forests have been lost to production and hundreds of species of wild animals have become rare or endangered.

Natural systems are not considered in terms of their ecological worth. They are viewed as a resource for man to exploit, often for the highest possible monetary gain. There is little interest in or knowledge of the ecological aspects of the biosphere—the fact that the entire system surrounding the Earth hangs together in a delicate web. And man is a part of the web.[6]

The biosphere is so immensely complicated that its workings are imperfectly understood. But it is known that any interactions of factors, however slight, can produce repercussions whose changes can span continents or even circle the Earth. To view life in a proper way, we cannot lose sight of the "unity of the biosphere" that lies beyond the plurality and essential rivalry of individual beings, especially the disturbance now being created by man. Of all the spheres of the Earth, the biosphere for us is the most central since it represents the "life-film" surrounding the Earth.

Many believe that the universe and our own Milky Way galaxy contain a large number of possibilities for life. But these are only speculations. We do know that life exists on the Earth, and it is the biosphere which has represented the growth of life for more than two billion years. It is the biosphere which today is undergoing serious ecological changes, the result of which we can only be partly certain. Rapid and global changes in the biosphere caused by human activities are viewed today as a serious threat to the future of all forms of life. We urgently need to devise ways to protect and preserve this thin veneer of life which encircles the globe.

Solidarity of Living Things

Many of us have grown up lacking a basic knowledge of the Earth's system and an understanding of how things evolved and how we as humans have emerged out of the matrix of life which now surrounds the globe. Too long we have lived with the myth that man has a different origin than other animals and that he stands outside of the natural interrelationships of the biosphere. Our lack of knowledge has been extremely costly as we observe the present deterioration of the environment.

The historic dichotomy which separates man from nature has been a primary source of the ecological crisis. It has led to the erroneous assumption that radical changes in the biosphere do not change conditions in the human environment. However, the social environment of man and the physical environment of nature do interrelate. Winston Churchill was aware of this continuous interplay between man and environment during the debate on restoring the House of Commons after World War II. Churchill feared that departure from the physical characteristics of the old House would affect the manner of debates and, as a result, the structure of English democracy. He stated, "We shape our buildings, and afterwards our buildings shape us."

All around us, from the depth of outer space to the minuteness of the atom, the universe holds together. Everything is dependent on everything else. All things are somehow linked together. This fact is well dramatized today through the dispersion of DDT to all parts of the globe, carried there through the delicate interconnections of the food chain. The entire cosmos with its billions upon billions of celestial bodies appears to have some sort of interrelatedness. Teilhard de Chardin writes, "The further and more deeply we penetrate into matter . . . the more we are confounded by the interdependence of its parts."[7] Again he writes, "Life was born and propagates itself on earth as a solitary pulsation."[8]

The living world is really like a giant tree of life whose branches have grown from one common trunk. The story of evolution is not simply a record of past events, it is going on now before our eyes. The trunk continues to issue new buds and branches. Man is the most recent and the most predominant new growth. His presence and activity overshadows all other activities. But the important lesson which must be learned is that although he may seem to be autonomous with respect to biospheric life, man still remains connected to the central evolutionary trunk. Lewis Mumford has written, "Though he is now the dominant species, his fate is still bound up with the prosperity of all forms of life." [9]

We must remove the false dichotomy that sets man and nature over against one another, not only as opposites, but as uncompromising enemies. Man's war with nature has been perpetuated century after century. Now it is apparent that nature is losing the battle. We are an integral part of the natural world and must realize that the global pollution of the environment is really an assault on our bodies and a threat to our own lives. Conrad Bonifizi asks, "How far dare man remove himself from nature and not be overcome or destroyed by it?" [10]

WHOLE EARTH THEOLOGY

5

The more we think of ourselves as masters of the Earth, the more we must be concerned with its prudent stewardship. —Bertrand de Jouvenel

The Hebrew concept of creation has been credited with destroying the magical view of nature held by other ancient religions. The Sumerian, Egyptian, and Babylonian religious systems held that their deity and man were a part of nature; nature was alive and man must pacify her before she could be approached. The Genesis account of creation is a declaration of the Hebrew faith that separates God from nature and distinguishes man as having a dominant position over the rest of creation. God was respected as fully autonomous and independent. Man and all created things were considered totally dependent on him.

Erich Fromm asserts that the Hebrew view of creation is fundamental to the evolution of humanness in man. Man is seen as being created in the image or likeness of God with a potential for continuous evolution of which the limits are not set. "Man the prisoner of nature," writes Fromm, "becomes free by becoming fully human." The aim of human action involves the constant process of "liberating oneself from the shackles that bind man to the past, to nature, to the clan, to idols." Fromm believes that man and nature must be enemies not to be reconciled until man has become fully human.[1]

A similar viewpoint is expressed by Eric Hoffer when he offers "a strategy for the war with nature."[2] He also points to the uniqueness of the ancient Hebrews whom he claims were the first to "enunciate a clear-cut separation between man and nature." According to Hoffer, the writers of the Old Testament chose as father of the race, not Esau, "a man of nature, whose garments, like those of Thoreau's ideal man smelled of grassy fields and flowery meadows," but his twin brother Jacob, who "preferred the inside of a tent to the great outdoors, and the smell of lentil soup to the smell of trees and fields." Hoffer argues that man's chief goal in life is still to become and stay human, and defend his achievements against the encroachment of nature.

As has been indicated earlier, Lynn White Jr. believes that it is this war-like attitude toward nature that Christianity inherited from Judaism which is at the root of our ecological crisis. White claims that Christianity not only established a dualism of man and nature but also insisted that it is God's will that man exploit nature for his proper ends. White writes, "By destroying pagan animism, Christianity made it possible to exploit nature in a mood of indifference to the feelings of natural objects."[3]

Harvey Cox admits that the biblical faith has caused a "disenchantment of nature" but goes on to state that modern man's attitude toward nature and the tendency to smash and brutalize it is perhaps a type of revenge of a "former prisoner against his captor," but this attitude is essentially "childish and is unquestionably a passing phase." Cox writes, "The mature secular man neither reverences nor ravages nature. His task is to tend it and make use of it, to assume the responsibility assigned to The Man, Adam." [4]

Man as Image

The popular Hebraic idea that man is made in the image of his creator has contributed significantly to an understanding of man's essential purpose on the Earth. Man as image, in the sense of likeness or resemblance, suggests that he is in some way a reflection of the creative spirit which has produced him. Interpreters generally agree that the image has less to do with physical or spiritual characteristics, and more to do with the particular task man is to perform. The argument for the close relationship between the image and man's task is based on an understanding of the Hebrew word, *selem,* which means an actual plastic work, a duplicate. Gerhard von Rad provides the following interpretive example:

> *Just as powerful earthly kings, to indicate their claim to dominion, erect an image of themselves in the provinces of their empire where they do not personally appear, so man is placed upon earth in God's image as God's sovereign emblem.*[5]

From the perspective of the Hebrew faith, it is clear that man's place on Earth has to do with authority over the nonhuman world. This authority does not proceed from man himself, but emerges from a source beyond human life. This *conferred* authority distinguishes man from all other creatures and endows him with characteristics which are shared by no other animal. Prominent among these are the abilities to engage in reflective thought and to communicate through diverse languages. His supremacy also enables him to organize intimate and well-functioning communities, and develop highly sophisticated cultures. Although man is a creature requiring food, rest, and security, he is also a kind of creator. He has the intrinsic qualities of bringing order out of chaos, of shaping meaningful objects from the Earth's substance, and of cultivating the land for food.

If man has been placed on the Earth as the *sovereign emblem* of the Creator, and if he is to reflect a mastery over the created order, then the "care of the earth" [6] becomes the fundamental human mandate. Man as the image of his creator places on him the special task of tending and caring for the whole of creation. Whether this injunction emerges from ancient religious sources, or becomes obvious from the biological differences between man and other creatures, or is evidenced historically in man's increasing control over nature, it is clear that man has an inevitable responsibility for his environment. Wherever he fails to care for his environment, he not only threatens his own existence, but is also in danger of destroying the image he bears, and his own identity as well. Regarding the meaning of this image, Dietrich Bonhoeffer writes, "This freedom of dominion directly includes our tie to the creatures who are ruled. The soil and the animals whose lord I am are the world in which I live, without which I am not." [7]

But the injunction to care for the earth is not always heeded. The exercise of dominion in the present century has wrought a devastating blow to nature and to its life-giving system. The rapine approach to the natural environment is gaining momentum, and the spoils which man carelessly deposits over the land rise up in a venomous assault on the human species. There is ample evidence that man has overplayed his dominant role, and the much-afflicted environment is preparing to strike back in order to restore the necessary equilibrium.

The forces which man is called on to control are no longer the traditional creatures of the natural order. New "demons" have emerged which are potentially far more destructive than the cataclysmic forces of raw nature. They are the result of man's indiscriminate exploitation of the natural world. They are the perilous consequence of an arrogant human enterprise. Man is making over large sectors of the land-mass, effecting radical environmental changes, the results of which will take decades to appraise. A revolution is taking place within man's environment which simultaneously threatens to poison the air shed over the Earth, strangle man's movement in the city, press humans into stressful spatial relationships, and harass citizens with a growing and persistent noise. The revolution is affecting all of us, and we are hard pressed to know what to do.

It is a strange coincidence that at the time when mankind has risen to a new plateau of confidence and control over nature, he finds himself surrounded by extensive environmental problems whose cause can be traced directly to human carelessness. At a point in human history when the "care of the earth" seems less a philosophic statement than a pragmatic reality, man has unwittingly dropped the ball.

Ecological Conscience

Lynn White Jr. suggests that the fundamental attitude toward nature and environment is based on a broadly accepted religious heritage. He writes, "Since the roots of our trouble are so largely religious, the remedy must also be essentially religious, whether we call it that or not." [8]

In the American tradition, which reflects Western thought, Judaism and Christianity have been primarily concerned with justice and compassion for man's treatment of man. They have traditionally assumed nature to be a mere backdrop for the human drama. It has been customary to interpret the Genesis command of "subdue" and "dominate" in a literal sense, providing the authoritative rationale for dominion over all of life. From this perspective, religion becomes the fashioner of human enterprise and serves as the catalytic agent enticing man to take over, change, use, and exploit.

Religion, while preoccupied with personal and social ethics, has avoided the vast area of man's treatment of nature. While religion has emphasized the other-worldly consequences of man's behavior in the social realm, it has not taught man to be concerned about the Earthly consequences in his treatment of nature. It has been left to that obscure science called "ecology" to arouse concern over boundaries and limits. The principle of balance in nature—the web of life, the necessity of controls over change—has stimulated scientists to plead for an "ecological conscience" in society.

The famous naturalist, Aldo Leopold, is credited with describing the process of "cooperation" within living organisms. Complex animals consisting of billions of cells are able to function only because of the "symbiotic" relationships among them. Each cell concedes a part of its autonomy toward the operation of the total organism. If it were not for this "willingness" to cooperate, the organism would remain a mass of inoperable cells not able to function as a unit.

Leopold was interested in the ethical content of this cooperation since it could be applied to the relationship of man to his environment. He was convinced that no ethic dealt with man's responsibility for the environment. But the extension of ethics to include man's relation to environment was, according to Leopold, "an evolutionary possibility and an ecological necessity." Leopold wrote:

> *All ethics so far evolved rest upon a single premise: that the individual is a member of a community of interdependent parts. His instincts prompt him to compete for his place in that community, but his ethics prompt him also to cooperate (perhaps in order that there be a place to compete for).*[9]

Leopold brought into clear focus one of the most important moral discoveries of our time, the "ecological conscience" centered in an awareness of man's true place as a dependent member of the life community. Leopold laid down a basic principle of the ecological conscience:

> *A thing is right when it tends to preserve the integrity, stability and beauty of the biotic community. It is wrong when it tends otherwise.*[10]

That the ecological conscience is not dominant in our society is quite obvious. It tends to be popular among conservationists and radical ecologists, but the majority of the population is only passively interested. Thus far, no ethic deals effectively with man's relation to his environment. The time has come when such an ethic will determine not only the future existence of man, but of other species which have lived on the planet for millions of years. Religion and ecology have been out of touch with each other for decades with no apparent catastrophic results. Continued severance, however, can only mean disaster. A reverence for the balance of nature must be incorporated into man's religious ideology, thereby molding human enterprise to correspond with ecological realities.

An application of the ancient mandate to have dominion over and subdue the Earth cannot be naively interpreted as license for exploitation of the natural environment. While the mandate may imply that man is superior, it also makes quite clear that man is responsible. Religion teaches that man is placed on the Earth as a steward who must care for his property. Ecology teaches that we are intimately connected to the natural order, and what we do and how we operate within that order will affect both the quality of our lives and our future destiny as a species. Religion in our day must review its entire understanding of nature and environment, and for the sake of mankind be open and responsive to the concerns and teachings of ecology. The human imperative to "care for the Earth" has never been as urgent as it is in the present time, and its fulfillment will depend, to a large extent, on the willingness and capability of religion and ecology to enter into meaningful dialogue.

Sisterhood of the Biosphere

On the surface it may appear that the religious community is ill-prepared to participate in the ecological movement since it has tended to epitomize man and depreciate nature. Conservationists, rather than theologians, have devoted themselves to the development of an ethic which relates man to his environment. But such an ethic has been slow to develop within the religious sector, particularly within Western religions. The time has now come, however, when such an ethic is the key to survival. Aldous Huxley asks: "Do we propose to live on this planet in symbiotic harmony with our environment? Or preferring to be wantonly stupid, shall we choose to live like suicidal parasites that kill their host and so destroy themselves?"[11]

The popular use of the word *nature* in our day refers to that part of the living substance which is nonhuman. We commonly make a separation between human and "subhuman" existence; yet such a separation is unknown in biblical thinking. The word *nature* has no equivalent in the Scriptures. The New Testament uses the word *creation* in place of nature to include both man and nonhuman reality. The ancient Hebrews did not advocate man's separation from the rest of creation, but were attempting to refute other polytheistic religions which held the view that nature and the "gods" were one and the same thing. Both Old and New Testaments assume a unity of man and nature, and emphasize the decisive role which man plays in the process of nature. The biblical doctrine of creation involves the unity in God of the multiplicity of entities in the world. Creation, according to the biblical teaching, is *all one piece*. This was assumed by Paul when he described the "whole creation" as participating in a groaning and longing until in the end it partakes of the glorious liberty of the children of God (Romans 8). "All things" (Colossians 1) hold together in a common, massive, fabric of creation totally dependent on God, the Creator of all that exists.

Not only do the biblical sources point to the "fatherhood" of God and the "brotherhood" of man, there is also an accompanying "sisterhood" which relates man to other living things. Joseph Sittler has stated: "In the Bible the earth is never called my mother, but she is called my sister. She is God's other creation who stands alongside me so there is a sibling relationship in his creation."[12] According to Sittler, "man is triply related to God from whom he comes; by the relationship of the fellowman, the other with whom he lives; and to the earth which is the garden of his ancient habitation." Sisterhood cannot be limited to the Earth alone, but must encompass the entire created order of planets, stars, and galaxies. But for our day a recognition of the "sisterhood of the biosphere" appears to be the dominant need.

In the last analysis, man does have dominion over all

the Earth, but this places him under grave obligations. No individual, society, or nation, has the right to over-utilize the world's resources for its own selfish interests or to maintain a particular living standard. It is admirable to be deeply committed to the well-being of the present generation of human beings who here and now inhabit the Earth. It is even more commendable for persons living today to become increasingly concerned about the future of their children and their children's children in the face of a worsening environment. But the new ethic of ecological responsibility must extend far beyond even this highly humanitarian concern. It includes the highest responsibility for the total, natural environment, the biosphere, and life itself, not human life only, but all of life in its varied and diverse forms. In the words of George Wald:

> *Our business is with life, not death. Our challenge is to give what account we can of what becomes of life in the solar system, this corner of the universe that is our home. . . .*[13]

Kenneth Boulding has written, ". . . the principle task of education in this day is to convey from one generation to the next a rich image of the total earth, that is, the idea of the earth as a total system."[14] The role of religion is similar, but it can do more. Religion can formulate a new environmental ethic that relates man to the Earth and emphasizes his responsibility for its preservation. A new ethic is needed that will transcend most of the values we have traditionally held concerning our world. The environmental ethic suggested by Aldo Leopold is a beginning point. But the new ethic goes beyond the treatment of land, plants, and animals. It encompasses the whole system of the Earth—the rivers, lands, oceans, biosphere, atmosphere, magnetosphere, and now also outer space. But this ethic also incorporates the rich Judeo-Christian heritage of an Earth that does not belong exclusively to man; he is entrusted to care for it, to "keep" and "till" it. Ownership of the Earth is not a part of man's privileged position with regard to the whole of creation. He is a steward and is expected to care for the garden which he *does not own.*

Guidelines for a Whole Earth Ethic

Several principles serve as components of the new ethic of the whole Earth. The first principle is that the Earth must be perceived as an infinitesimal planet in a massive universe. We are travelers together on a tiny spacecraft dependent on one another, dependent on the plants and animals that live here, dependent on clear air and water, and dependent on the energy that comes from the sun. However threatening nature appears to be, from the perspective of the cosmic universe it is fragile and kindly, a benign habitat for man.

The second principle relates to the age of the Earth. Our planet is indeed ancient, but man was only recently born. He has grown out of the matrix of life extending over several billion years. Man is the freshest, the most advanced form of life in the solar system and perhaps in the universe. But his future evolution holds the greatest potential. The major part of his life as a species lies in the future. And for much of that future he will be confined to the Earth's environment. We must hereafter appraise the threats of pollution in light of the latent possibilities for man's future.

The third principle is that we learn to treat the Earth as a total system of air, water, and land. We can no longer isolate state, national, or continental boundaries; all are intimately connected parts of a vast system of concentric spheres. The pollution emerging from one city becomes the common pollution of all cities when mixed with the gaseous effluents being poured into the air shed. The dumping of chemical waste in one river system has the effect of polluting the general quality of the hydrosphere. Everyone's environment merges into everyone else's. We can no longer treat the problems of environment in piecemeal fashion; our efforts must be intercommunal, interstate, and international in scope.

The fourth principle relates to the ecological balance particularly of the biosphere where all life is enclosed. We must consider the essential equilibrium of this sphere before initiating any action that would destroy the ecosystem. This balance is being threatened today by the multiple forms of urban growth, technological innovation, and the expansion of industries throughout the world. These disturbances pose the threat of destroying complex genetic substance which, once destroyed, cannot be recreated. As a major "simplifier" of ecosystems, man causes a greater vulnerability of living things to disease. A healthy environment for plants and animals, and also for man, is one that encourages diversity and allows for natural balance.

The fifth principle of the whole Earth ethic requires us to recognize that we live in a finite world. Our resources are not limitless, and when those that are nonrenewable are consumed, they cannot be replenished. Our present resources should be carefully husbanded and conserved. The physical frontiers no longer exist, and hereafter we must learn to produce and consume less. We are responsible for the exploding population in the Earth today, but we must also concern ourselves with the needs of future generations who will need oil, coal, iron, and other resources.

The sixth principle of the whole Earth ethic is the most urgent. We must learn to control human fertility. Nothing threatens life on the planet more dramatically than does the growth in human numbers. It is absolutely imperative that the human birth rate be curtailed. Man's superior intelligence does not entitle him to convert the substance of the biosphere into a mass of humanity. Every effort must be made to encourage voluntary birth control in all countries. However, if individuals are unable to discipline themselves for the sake of the Earth, mandatory controls must be enforced so that some degree of quality can be maintained in the human species.

The seventh principle relates to the traditional freedom given to technology. We need to reject the premise that technology alone can provide answers to all or most of our ecological problems. Technology has a tendency to create the need for more technology just as now we see a new growth in anti-pollution industries. We must hereafter force technology, or those who direct technology, to serve the essential needs of man, including his need for a quality environment.

The eighth and final principle is concerned with ownership. The whole Earth ethic rejects the assumption that man *owns* the Earth. The Judeo-Christian imagery of man as a "steward," "gardener," or "husbandman" has practical value for our day. The air shed, the river systems, the oceans, and the forests do not belong to us. We have inherited our present environment from past generations, and we will transmit what we have to future generations. We have been great exploiters, but poor stewards. The actions of those yet living today have brought us to the brink of disaster. It is left to the rest of us to reassign priorities so that the inheritance might be passed on in at least as good condition as it was when we received it.

Many argue today for a new style of life for all citizens. The productive and consumptive patterns of society must be replaced by an economics of stability, or on a global scale, an economics of scarcity. But more basic is the need for a new *style of thought*. What is needed is a new understanding of man's place in the universe: a new vision of his responsibility for the Earth, a new appreciation for being at the end of a long evolutionary process, a new excitement regarding the future, and a new determination to keep this planet from becoming uninhabitable. Man needs a new awareness, a new consciousness, a new passion for the *whole Earth*.

The environmental crisis signals both a potential end to life on the planet, and the beginning of a new human consciousness. The present crisis requires man to view his destiny as coincidental with the destiny of the entire Earth. Human decisions must increasingly be measured in terms of their impact on Earth systems.

Man's role as steward has never been more urgent. His concept of stewardship has now expanded to encompass the *whole Earth* rather than individual parts. Our theology must take on a new holistic view of man in the web of life; and life as a delicate veneer spread over the surface of planet Earth; and Earth as a tiny ship floating in a massive sea of almost endless space. A theology of the whole Earth leads man to formulate an appropriate *thought style* so that he may accept the new responsibilities placed on him by the advent of *spaceship Earth*.

The religious community, now committed to the personal needs of humanity, must be concerned about the environment of man, ecology, and the life systems on Earth. The forces at work to destroy the thin, green film of life that encircles the globe have gained a strangler's hold, and the forebodings by scientific leaders should suffice to awaken all of us to a whole new dimension of service. This service will require a thorough review of the traditional understanding of stewardship. But more important, it will require that we perceive the Earth in a new way. The Earth is first and foremost a gift to be received with joy. Our first responsibility is to celebrate the Earth as the "garden" in which man is placed to live out his days. The Earth is our sister with whom we share "creaturehood." And if she is our sister, we can no longer treat her as a mere *thing*. "If we hope to be well treated by nature," writes Aldous Huxley, "we must stop talking about 'mere things' and start treating our planet with intelligence and consideration." [15]

The present ecologic crisis will not be dealt with through the quietistic efforts of preservationists, nor will a kindly sermon on Rogation Sunday be sufficient. It is time to take a stand on behalf of the Earth itself. The religious community must finally reckon with what has been true from the beginning: the creative and redemptive processes encompass not only man, but *all things*—from the smallest atom to the largest galaxy. Creation is being driven toward an essential unification of all its parts. Nothing stands outside the scope of the redemptive process; not plants, not animals, not lakes, not forests, not polluted streams. In the battle to save the environment the religious community should become a primary advocate. It can lead the way to teach mankind to *celebrate* the Earth in its totality, and keep its many treasures for future generations to enjoy.

ECO-CELEBRATION

6

You'll conquer the present
suspiciously fast
if you smell of the future
—and stink of the past.

—Piet Hein

Students of theology who translate the New Testament from its original Greek are familiar with *oikos,* the root word for ecology. They ordinarily translate the word as "house" or "household" referring to home or family unit. But the Greek phrase which is most similar to ecology is *ta panta,* meaning "all things." When the early church spoke of the extent of redemption it used *ta panta.* The scope of redemption includes the totality of all that is. Nothing stands outside the redemptive process.

In a similar way ecology provides the extent of natural balance. The scope of ecology includes "everything" in the biotic community. "Everything" in the ecosystem is related to everything else. Ecology's main contribution to the present crisis in environment is its view of the whole and the interrelationship of all its parts. Nothing stands outside the basic equilibrium of the Earth's systems.

There are various uses of ecology's prefix "eco." Paul Ehrlich refers to "eco-catastrophe" in a *Ramparts* magazine article.[1] Through a scenario he describes what our world will be like in ten years if the present course of environmental destruction is allowed to continue. "Eco-catastrophe" is coined to describe the most spectacular of man's attacks on the systems that sustain his life. The Sierra Club has produced a book titled *Ecotactics,* the science of arranging and maneuvering all available forces in action against enemies of the Earth.[2] A film produced by Luther Gerlach bears the title, "People Eco-Action," describing a new social movement in ecology that threatens to be the most devisive force in our nation.[3] Thomas Turner refers to "eco-pornography," which is the "pornography" of advertising agencies who produce subtle television and magazine commercials for major industrial polluters.[4] The uses of "eco" imply that ecology is concerned with social action. The term is being used to distinguish preservationists from exploiters, activists from the inactive, environmentalists from technologists, the "friends of the Earth" from the "enemies of the Earth." Ecology has come to mean a negative, doomsday view of the future and it is being used to divide rather than to unite. Even the ecology flag bears the Greek letter *Theta* to warn of death.

Ecology is neither activistic nor divisive. Its basic meaning is that everything is linked to everything else. Life on the planet holds together and displays a continuous interdependence. Ecology places restraints on overproduction, overconsumption, overefficiency, and overreproduction. It is more interested in the whole rather than individual parts. Its message to us is not one of doom but one of hope. It appeals to the "landscape of our minds" and the "wholeness of our spirits." Ecology is a science that encompasses all sciences; it is the "art" of the sciences and the "science" of the arts. It is the one discipline with categories broad enough to unite us, and the necessary flexibility to give us proper freedom.

The Orgy Must End

The principles of ecology are embarrassingly lacking in American thinking. We have operated and continue to operate as though the physical frontiers were sufficient to serve our increasing demands. The American spirit is based on the principle of exploitation. French historian, de Tocqueville, found it difficult to describe the forcefulness with which American settlers pushed the Indians out of the way and seized new lands. He wrote, "Here is a civilized people who are also the most grasping nation on the globe."

Paul Ehrlich identifies this nation's unquenchable hunger for the resources of the Earth when he states that although we represent six percent of the world's population, ". . . the United States uses well over half of all the raw materials consumed each year."[5] An American child is 50 times more burdensome on the environment than is a child from India. Barry Commoner has indicated that the United States produces 11,000 calories per capita per day when each person requires an average of only 2,500. The same overindulgence is reflected in the proliferation of highways, the production of millions of automobiles each year, and the disposal of monumental piles of waste—100 million tires, 20 million tons of paper, 28 million bottles, and 48 billion cans.

While the United States produces and consumes and enjoys the pleasures of affluence, nearly two-thirds of the world's population does not receive adequate daily nutrition and millions are threatened with starvation. Dutch theologian, J. C. Hoekendijk has stated, "Solidarity with this world can be nothing else than solidarity in poverty."[6] If all the resources of the Earth were equally shared with its 3.5 billion inhabitants, we would all be among the poor.

Georg Borgstrom, author of *The Hungry Planet,* has stated that if all the food in the world were equally distributed and each person received identical quantities, we would all be malnourished. If the world's food supply were distributed at the present U.S. dietary level, it would feed only about one-third of the human race. Over two billion of the world's 3.5 billion people experience food and water shortages. In sharp contrast to their misery, a "luxury club" of nearly 400 million people, half of them Americans, enjoy a rich and steadily more abundant diet as well as a high standard of living.

If the world's population were compressed into a single town of 1,000 persons, 60 residents would be American. This six percent would receive half of the total income. The 60 Americans, as a result of their utilization of the town's available resources, would have 15 times as many possessions per person compared to the remaining 940. The American average life expectancy would be 70 years compared to about 40 years for the remainder. The lowest income group among the Americans, including a small percent not free from hunger, would be better off by far than the average among non-Americans. The United States is systematically plundering the planet's physical resources. We might well ask: "Is the United States a planetary disease?"[7]

The American spirit and the ecologic crisis are so closely related that even on a global scale we cannot treat them as separate. The solution to the crisis is inextricably bound to the image we Americans have of ourselves. The response of this nation to the demands of ecology will determine to a great extent whether mankind can go on. It brings the crisis in environment frighteningly close to home.

We must learn that the present standard of living in this country is made possible only through the utilization of resources from other nations. As the supply of such resources as natural gas, oil, and coal become depleted, the large industrial corporations must seek new reserves in other lands and along other shores. We must ask ourselves whether this is humane and just. What about future needs of developing nations? Will there be an ample supply for these countries with their burgeoning populations and newly acquired technologies? We cannot think of our nation apart from other nations; we must be conscious of the deteriorating environments of other lands and of the damage to the entire Earth.

We are fascinated by the gimmicks and the gadgetry of technology and we are in a rush to exploit every last resource. We fail to count the consequences. Our rivers and lakes are open sewers; emphysema is claiming lives at an alarming rate; noise is damaging our sense of hearing and filling us with anxiety. We suicidally abuse our environment—by our numbers, our affluence, and our technology. We are experiencing the squeeze of too many people and too many things. We feel it in power failures and water shortages; in smog and sewage and slums. By the year 2000 at our present rate of growth we must make room for 300 million Americans, 244 million cars, 6 trillion pounds of garbage.

"The land has come to be treated as a huge, two-dimensional biochemical factory powered by sunlight," writes Nigel Calder.[8] "We behave often," states René Dubos, "as if we were the last generation to inhabit the earth."[9] Unless we can limit our numbers—unless we recognize that the Earth does not belong to us, but we belong to it—man may become a member of the endangered species. Not only are the destruction of natural beauty or the sacrifice of recreational areas crimes in themselves, but the major crime is our interference with the entire ecology, with the balance of nature on which persistence of life on this planet depends.

William Vogt, author of *Road to Survival,* once stated that a man with ten children has the wrong image to be president of the United States. Therefore, Vogt would not support the candidacy of Robert Kennedy. He was not addressing himself to the leadership capabilities of Kennedy, but merely pointing out that "children of the affluent" are the greatest hazard not only to the American environment but ultimately to the entire Earth system. There is a popular assumption that the poor of the Earth are the greatest polluters. But the children of the affluent make the biggest demands on the Earth's resources; they are the biggest polluters. Controlling the poor is not the problem, but controlling the rich. And in terms of the world's people, most Americans are among the rich.

Americans are participating in an *orgy* of their own making. The glut of things produced and consumed each day resembles an obese man gorging himself at the dinner table. We produce and consume much more than we need. Our trade centers stink of overindulgence. There are too many things to whet our enormous appetites, and too many messages that seductively shout—buy! buy! buy! Everything is made to be momentarily disposed of. Nothing is made to last. One automobile per family in Latin America would be a luxury. But in America two, three, four, and five cars per family unit are not uncommon. One television set per family unit in India would strain their industrial production, but in America it is no longer uncommon to have two and three sets in an average home. We have so many things—and soon we will have more. Our greed marks us! We are quickly becoming parasites eating away at the Earth's resources. We are, indeed, a planetary disease. And we must be stopped. We stand now as the one greatest threat to the future of the planet, and the future of life on the planet.

The film "Junkdump" describes our predicament.[10] In a satirical vein the film describes two persons, a husband and wife, living in a city dump. The husband works at a second dump where he sells used parts to various buyers. His day lingers on—a steady humdrum of the exchange of discarded things. The husband returns home, greets his wife, and holds the baby. The baby is a discarded doll—a product of the dump. But mom and dad don't know the difference; the real and unreal have blended indistinguishably together. Flashbacks remind us that the thousands upon thousands of autos composing the morning rush hour will soon become monumental piles of waste. The only difference between the mountains of new merchandise found in discount stores across our land and the local city dump is a few months of use. The faster things turn to junk, the more our technologies produce. The orgy has become a sickness, a social disease. It has begun to destroy the mind and heart of America.

Alternatives to Gluttony

Only three alternatives are open to us. We can continue in our frontier mentality as though the Earth's resources and reservoirs for waste were infinite, hoping that somehow an unknown force will rescue us from what certainly will be a catastrophic end. Or we can choose to return to the past by rejecting our high level technology and our present standard of living. Or we can accept the rubrics of the whole Earth ethic by changing our individual life styles and societal priorities.

Although the hippie movement may not be an outgrowth of conservation or ecology, it does suggest a possible life style for the future. The hippie, whatever else we may say about him, is much less destructive of the environment than a member of the affluent society. This is not to suggest that we must choose the hippie as a life model. It simply indicates that it will become necessary, given certain ecological realities, to define quality of life and standard of living in different terms.

It appears that a large percentage of young people will take the Earth more seriously than did their parents and grandparents. They will be prepared to make "radical personal sacrifices" for the sake of future life on the planet. Some experimentation is taking place among the young in various forms of communes. Some experiments are done for ecological reasons. Pesticides and herbicides are not used. Human wastes are disposed of by returning them carefully to the soil. Members consume only organic foods. But to assume that an entire urban society can live this way would be naive. Our population, our governments, our educational structures, our transportation and communication systems are far too complex and must be maintained at a sufficient level to prevent the total collapse of society. Asceticism is not the answer. The few who seek the austere life should be encouraged to do so. For the vast majority something more formal must be offered.

We cannot continue with the present reckless, exploitative, romantic, and violent behavior. Kenneth Boulding refers to this as the "open" or "cowboy economy" as compared with the "closed" economy of the future which he calls the "spaceman economy."[11] The cowboy economy is based on how much can be consumed and produced. The spaceman economy minimizes consumption and production and places major emphasis on the nature, extent, quality, and complexity of the total capital stock, including the state of the human bodies and minds included in the system. The spaceman economy will require taking on the rubrics of the whole Earth ethic.

The new ethic of the Earth will require major adjustments in our value system. Pollution of the environment can be traced directly to the commonly accepted patterns of living practiced by citizens everywhere. The choice of a home cleansing agent, the decision to purchase a second automobile, the ownership of a powered boat, the size of one's family, the use of air-conditioners and dishwashers, the preoccupation with fashion and dress, the construction of a second home—all have an effect on the environment. It has become far too convenient to point an accusing finger at industry, with no appraisal of what one individual contributes to the ecological crisis by the way he lives. Too few persons are able to see the close relationship between the electrical outlet on their living room wall and the various forms of pollution caused by power generation. In most cases, what we call pollution of the environment can be traced back to the life styles of our society. And most of us are reluctant to change, in spite of the warnings that come from scientists and government officials.

The reassignment of societal priorities will require certain exchanges or trade-offs if we are to achieve our goal of a quality environment. We must assess the need of providing food for the growing numbers of people against the measurable depletion of the soil. The automobile has been cited as the main polluter of air and a hazard to life in the large cities. It is necessary to assess the jobs of thousands of auto workers and the lack of public mass transit against the frustration of traffic congestion, the increase of air pollution, and the accumulation of millions of discarded cars. A small taconite community along the shore of Lake Superior is wholely dependent on the one existing mining industry. The survival of this small community must be assessed against the further pollution of a great lake. We must assess the individual freedom of those who want large families against the demand for population control; the elimination of crop insects and malaria against the contamination of the soil through DDT; the privilege of unrestricted electrical consumption against the growing web of nuclear power plants throughout the nation.

It may be traumatic for many of us to adopt a less wasteful style of living. People can no longer accuse greedy businessmen or mindless industry for their environmental mistakes. Individuals must fight their own instincts to spend and to consume. This will not come easily. The young will make the transition more easily than the old. The old are difficult to convince because they do not want to go back to the inconvenience of the past. The frontier mentality of the old must die with them. We are moving into a new age of "cooperation toward survival." It reminds us of the environmental ethic of Aldo Leopold: one's instincts prompt him to compete, but his ethics prompt him to cooperate, perhaps so there will be a *place* to compete for.

An Ecology of Joy

Several prominent scientists have refuted the popular assumption that the solution to the ecologic crisis is technical. By this they mean that the crisis will not be solved simply by developing new technologies, allocating more monies for abatement, or by passing new and stricter laws. The problem is not at its basis technical, legal, or economic. This implies that the problem lies deeper within the fabric of our culture; it is a part of our system of social values, or perhaps more basic, it is related closely to the nature of man. The crisis is a product of human nature and therefore it requires a moral and religious solution. In short, the ecologic crisis is a deeply human problem and must be solved through a change of mind and heart.

Man relates to the Earth in two ways. He perceives it

through *technique,* and he learns to appreciate it through *aesthetic* experiences. In our day it is obvious that technique has played the dominant role. Whenever modern man thinks of the Earth his mind begins to "estimate" the economic profit from the sale of a plot of land, or "calculate" the crop production of a fertile farm, or "design" the plan for a massive housing project. The Earth is seen as a great resource to be mined, dredged, excavated, tiled, or chopped down. Swamps, rivers, lakes, forests, plains are all made to serve the purpose of man's grand scheme.

Although *technique* dominates, man still retains a great reserve of *aesthetic* appreciation for the Earth's beauty and magnetic appeal. The preservation of national parks and wild rivers, the protection of threatened wildlife species, and the growing interest in outdoor recreation testify that the Earth continues to appeal to man's sense of wonder and beauty. For thousands of years the Earth has not only been the source of his sustenance and the object of his toil, it has also been the basis for much of his joy. Yet, technique continues to rule. And the Earth suffers. Joseph Sittler writes, "If the creation, including our fellow creatures, is impiously used apart from a gracious primeval joy in it, the very richness of the creation becomes a judgment." [12]

Creation in our day has indeed become a judgment. The ecological crisis is an overt symptom of our impious use. We must now learn to celebrate the Earth, for through celebration we grow in appreciation of it as a treasury of good things. The present orgy is certainly not celebration; its efforts are to diminish and destroy. Celebration means to "keep" and to "preserve" so that it can be enjoyed again and again. We can celebrate the Earth in various ways and thereby keep it for ourselves, our children, and the multitudes of generations yet unborn.

We celebrate the Earth by enjoying a slower *tempo of life*. We are caught in a whirlwind of motion that leads ultimately to chaos. Automobiles, jet planes, and our own daily lives are geared to speed. Because we are in such a rush, we burden our environment with freeways, expressways, high powered engines, traffic snarls, noise, confusion, and human bodies ridden with anxiety. The time has come to slow the pace of society. Mahatma Gandhi once said, "There's more to life than increasing its speed." We must resist our insanity for "speed at any price." Celebration is ruined by new technologies that promise to take us to our destination 30 minutes faster, but in the process reduce the quality of environment for other citizens. We can discover again the joy of walking, of bicycling, of hiking. Reducing the tempo of life will teach us to cherish the smell of fresh air, the sound of birds, the fascination of the city, and the contentment of the countryside. A slower pace for society will relieve part of the burden speed places on the environment.

We celebrate the Earth by learning to enjoy the quality of *smallness*. We have become worshippers of bigness: larger cars, larger billboards, larger smokestacks, larger factories, larger machines, larger budgets, and a larger gross national product. Our motto has been: bigger is always better. But the preoccupation with quantity has substantially reduced the quality of our environment. We can refute the assumption that bigger is better, or that growth in itself leads to quality by giving greater attention to the virtues of smallness. In a limited, closed environment, smallness is obviously an asset. We celebrate the Earth by enjoying controlled size: small families, small populations, small cities, small automobiles, small industries. We celebrate by enjoying the challenge to miniaturize new technology and stabilize economies. In our increasingly crowded society we celebrate by making the most out of the least.

We celebrate the Earth by placing a high value on the importance of *open space*. As our society becomes increasingly more urban, crowding becomes more intense. Our densities are leading to a reduction in the quality of the environment. High densities require highways, streets, parking lots, industrial sites, shopping centers, housing projects, utility lines, auditoriums, schools, colleges, and churches. Space in an urban setting is being sold at a premium. Nearly every available open space is taken for residential, commercial, or civic use. Even the space over freeways is considered a salable commodity. This rush for space results in low priorities for open space; for parks, playgrounds, hiking trails, horse trails, and open corridors of greenery surrounding metropolitan areas. Open space is one of the greatest assets in an urban area. Our celebrations should not be limited to the newest and largest housing developments, factories, and office buildings. Celebrating the Earth means exuberant enjoyment of undeveloped, open land.

We celebrate the Earth by restoring *quiet* to the environment. Reports state that the decibels of sound in the public environment are slowly rising. A report of the World Health Organization indicates that noise has become a menace, "the mental pollution it causes is becoming more pernicious than even air and water pollution." Jet planes, loudspeakers, advertising commercials, transistor radios, motorbikes, buses, trucks, auto horns, and "Muzak" all contribute to the din at home, at work, in the skies, and streets. Noise contributes to nervous disease, insomnia, tension, and accidents. Offensive and persistent noise will continue to increase as our cities grow. The public environment becomes a receptacle for everyone's "sounds." Quiet is desperately needed and adds substantially to the quality of environment.

We celebrate the Earth by seeking *beauty* in nature and in design. Much of our environment represents stark ugliness. *God's Own Junkyard* by Peter Blake reveals the degree to which beauty is either nonexistent or being destroyed. Eric Sevareid writes, "With breathtaking rapidity we are destroying all that was lovely to look at and are turning America into a prison house of the spirit." The franchise architecture of service stations, restaurant chains, and discount stores turn urban environments into communities of sameness. The city is marked by massive billboards, streets lined with autos, utility structures, and unkept buildings. Natural beauty is scarred with auto junkyards, housing projects, and strip mining. Developers of residential and commercial property care little about preserving the natural environment. We celebrate by sharing our appreciation of beauty and insisting that beauty be part of a quality environment.

We celebrate the Earth by pursuing true *leisure*. What

is typically referred to as leisure (discretionary time) is increasingly becoming another form of work. Rather than productive work, it is "consumptive" work. The typical weekend "escape" from crowded cities to distant lakeside cabins or winter resorts becomes one additional burden for our environment. It results in more traffic, more litter, more pollution, and more accidents. Men in the pursuit of leisure spend entire weekends driving high-powered boats, snowmobiles, or motorcycles in recreation areas. Consumptive leisure threatens to become a new major hazard to the environment. New motorized vehicles which will travel on land, water, ice, snow, mud, swamps, and thickets will make the wilderness resemble the littered countryside. We celebrate by enjoying simple forms of leisure

We celebrate the Earth by practicing *moderation.* Joseph Wood Krutch once wrote, "Security depends not so much upon how much you have as upon how much you can do without." We are great exploiters of the Earth's resources and, as a result, we are not secure as a nation. Our vital resources such as oil and gas become more scarce each year. Our capacity to consume on a worldwide scale will eventually make the entire Earth vulnerable in terms of fuel resources. Our greed adds little to the quality of our lives because it reduces critically the quality of environment. Our future security and the security of the entire world depend on a major shift in our use of resources. Moderation in the use of resources is a form of celebrating the Earth.

We celebrate the Earth when we control *procreation* in our own species. Procreation is intrinsic to the process of creation. But when the population explodes so that creation itself is threatened, procreation becomes a form of blasphemy. Individual couples can no longer bear offspring indiscriminately. Too many people place a great burden on natural systems to produce sufficient food and absorb massive quantities of waste. Voluntary family planning does not give individuals the right to have as many children as they want. Each human celebrates the Earth as he plans to control procreation within

new problems of which we are not aware today. The major transition we must pass through is learning to live within the confines of a finite environment. We will learn individually and collectively to live creatively within the boundaries of ecology. But in so doing, we will discover a new spectrum of frontiers far more challenging to human ingenuity than were the frontiers of the American wilderness.

NOTES

PREFACE

1. William Paddock and Paul Paddock, *Famine—1975!* (Boston, Toronto: Little, Brown & Co., 1967).
2. Paul R. Ehrlich, *The Population Bomb* (New York: Ballantine Books, 1969).
3. John R. Platt, "What We Must Do," *Science,* vol. 166, no. 3909, November 28, 1969, p. 1115.

CHAPTER 1

1. Frank M. Potter Jr., "Everyone Wants to Save the Environment But No One Knows Quite What to Do," *Center Magazine,* vol. III, no. 2, March 1970.
2. "Intergovernmental Conference of Experts on the Scientific Basis for Rational Use and Conservation of the Resources of the Biosphere," UNESCO, Paris, September 4-13, 1968.
3. "UN Votes to Hold Conference on the Human Environment in 1972," *The Conservation Foundation Letter,* February 24, 1969, p. 2.
4. René Dubos, "We Can't Buy Our Way Out," *Psychology Today,* March 1970, p. 87.
5. Peter Schrag, "Life on a Dying Lake," *Saturday Review,* September 20, 1969, pp. 19-20.
6. "Ecology: Menace in the Skies," *Time Magazine,* vol. 89, no. 4, January 27, 1967, p. 52.
7. "The Unforeseen International Ecologic Boomerang," *Natural History Special Supplement* based on the conference, The Ecological Aspects of International Development Programs jointly arranged by the Conservation Foundation, Washington, D.C., and the Center for the Biology of Natural Systems of Washington University, St. Louis, Missouri, held in Warrenton, Virginia, December 9-11, 1968.
8. Garrett Hardin, "The Tragedy of the Commons," *Science,* vol. 162, December 13, 1968, pp. 1243-1248.
9. Thomas Merton. "The Wild Places," *Center Magazine,* vol. 1, no. 5, July 1968, pp. 40-44.
10. René Dubos, "Man and His Environment: Adaptations and Interactions," *The Fitness of Man's Environment* (Smithsonian Annual II), (Washington, D.C.: Smithsonian Institution Press, 1968), p. 243-244.
11. Lynn White Jr., "The Historical Roots of Our Ecologic Crisis," *Science,* vol. 155, no. 3767, March 10, 1967, pp. 1203-1207.
12. "Ecology: A Cause Becomes a Mass Movement," *Life Magazine,* vol. 68, no. 3, January 30, 1970, p. 23.
13. Quoted by Pierre Bertaux, "The Future of Man," *Environment and Change: The Next Fifty Years,* ed. William R. Ewald, Jr. (Bloomington: Indiana University Press, 1968), p. 19.

14. See "Toward Balanced Growth: Quantity with Quality" (Report of the National Goals Research Staff, July 4, 1970. Available from Government Printing Office, Washington, D.C. 20402). For background on the NGRS, see "The White House Looks to the Future," *The Futurist,* August 1969.
15. G. B. Gunlogson, *Redistribution of People: This Is the Greatest Challenge Facing America Today* (St. Joseph, Michigan: The American Society of Agricultural Engineers), p. 1.
16. Lord Ritchie-Calder, "Polluting the Environment," *The Center Magazine,* vol. II, no. 3, May 1969, p. 7.
17. René Dubos, *So Human an Animal* (New York: Charles Scribner's Sons, 1968), p. 144.
18. Excerpts from a speech by James Russell Wiggins presented at the University of Minnesota during its fall commencement in 1968, *Minneapolis Tribune,* December 15, 1968.

CHAPTER 2

1. Rachel Carson, *Silent Spring* (Greenwich, Conn.: Fawcett Crest Books, 1962).
2. Stewart L. Udall, *The Quiet Crisis* (New York: Avon Book Division, The Hearst Corporation, 1963).
3. Robert Ardrey, *African Genesis* (New York: Dell Publishing Co., Laurel Ed., 1961), pp. 146-147.
4. Ritchie Calder, *Man and Cosmos* (New York: New American Library, Mentor Books, 1968), p. 47.
5. Arthur Beiser and the editors of *Life, The Earth* (New York: Time, Inc., 1963), p. 10.
6. Pierre Teilhard de Chardin, *Building the Earth* (New York: Avon Book Division, The Hearst Corporation, 1969), p. 17.
7. Phillip D. Stern, *Our Space Environment* (New York: Holt, Rinehart and Winston, 1966), p. 97.
8. David Bergamini and the editors of *Life, The Universe* (New York: Time, Inc., 1966), p. 97.
9. Beiser, p. 12.
10. William T. Skilling and Robert S. Richardson, *A Brief Text in Astronomy* (New York: Henry Holt & Co., 1954), p. 38.
11. Isaac Asimov, *To the Ends of the Universe* (New York: Walker & Co., 1967), p. 28.
12. Bergamini, pp. 153-54.
13. S.P.R. Charter, *Man on Earth* (Sausalito, Calif.: Contact Editions, 1962), p. 199.
14. R. Buckminster Fuller, *Operating Manual for Spaceship Earth* (Carbondale, Ill.: Southern Illinois University Press, 1969).
15. Adlai Stevenson, a speech given before the Economic and Social Council, Geneva, Switzerland, July 9, 1965.
16. Archibald MacLeish, *New York Times Service,* December, 1968.

CHAPTER 3

1. Stephen H. Dole, "Probability of Occurrence of Habitable Planets," *Habitable Planets for Man* (New York: Blaisdell Publishing Co., 1964), pp. 82-105.
2. Arther Beiser and the editors of *Life, The Earth* (New York: Time, Inc., 1962), p. 35.
3. Ruth Moore and the editors of *Life, Evolution* (New York: Time, Inc., 1962), p. 9.
4. Beiser, p. 132.
5. *Ibid.*
6. Theodosius Dobzhansky, *The Biology of Ultimate Concern* (New York: The New American Library, 1967), p. 33.
7. John N. Deely, "The Vision of Man in Teilhard de Chardin," *Listening,* Autumn 1966, p. 203.
8. Pierre Teilhard de Chardin, *The Phenomenon of Man* (New York: Harper & Row, 1959), p. 189.
9. *Ibid.,* p. 224.
10. Georges Crespy, *From Science to Theology,* trans. George H. Shriver (Nashville: Abingdon Press, 1968), p. 166.

CHAPTER 4

1. John Lear, "The Pulse of the Earth," *Saturday Review,* February 1, 1968, pp. 49-52.
2. Pierre Teilhard de Chardin, *The Phenomenon of Man* (New York: Harper & Row, 1969), pp. 96-97.
3. *Ibid.,* p. 97.
4. *Ibid.,* pp. 180-84.
5. Kenneth E. Boulding, "Ecology & Environment," *Trans-Action,* March 1970, p. 38.
6. See John H. Storer, *Man in the Web of Life* (New York & Toronto: The New American Library, 1968).
7. Teilhard de Chardin, *The Phenomenon of Man,* p. 44.
8. *Ibid.,* p. 102.
9. Lewis Mumford, *The Transformations of Man* (New York: Collier Books, 1956), p. 24.
10. Conrad Bonifazi, *A Theology of Things* (Philadelphia & New York: J. B. Lippincott Co., 1967), p. 26.

CHAPTER 5

1. Erich Fromm, *You Shall Be as Gods* (New York: Holt, Rinehart & Winston, 1966), p. 70.
2. Eric Hoffer, "A Strategy for the War with Nature," *Saturday Review,* February 5, 1969, p. 29.
3. Lynn White, Jr., "The Historical Roots of Our Ecologic Crisis," *Science,* vol. 155, no. 3767, p. 1205.
4. Harvey Cox, *The Secular City* (New York: The Macmillan Co., 1965), p. 23.
5. Gerhard von Rad, *Genesis,* trans. John H. Marks (Philadelphia: The Westminster Press, 1961), p. 58.

6. See Joseph Sittler, "The Care of the Earth," *The Care of the Earth and Other University Sermons* (Philadelphia: Fortress Press, 1964), pp. 88-98.
7. Dietrich Bonhoeffer, *Creation and Fall* (London: SCM Press Ltd., 1969), p. 37.
8. White, p. 1207.
9. Aldo Leopold, *A Sand County Almanac* (New York: Oxford University Press, 1966), p. 219.
10. *Ibid.,* p. 240.
11. Aldous Huxley, "The Politics of Ecology," An Occasional Paper on the Free Society published by the Center for the Study of Democratic Institutions, 1963, p. 6.
12. A lecture delivered at the Augsburg College symposium entitled, "Man and His Earth: Reaching for Quality Environment," February 4, 1969, Minneapolis, Minnesota.
13. Dr. George Wald, addressing students and professors at the Massachusetts Institute of Technology, March 4, 1969.
14. Kenneth E. Boulding, "Ecology & Environment," *Trans-Action,* March 1970, p. 38.
15. Huxley, p. 6.

CHAPTER 6

1. Paul R. Ehrlich, "Eco-Catastrophe!" *Ramparts,* September 1969.
2. John G. Mitchell and Constance L. Stallings, eds., *Ecotactics: The Sierra Club Handbook for Environment Activists* (New York: Pocket Books, 1970).
3. Luther P. Gerlach, "People Eco-Action," a documentary film about the mushrooming movement to save the environment. Minneapolis, Minn.: University of Minnesota, Department of Audiovisual Extension, 1970.
4. Thomas Turner, "Eco-Pornography or How to Spot an Ecological Phony," *The Environmental Handbook,* ed. Garrett De Bell (New York: Ballantine Books, 1970), pp. 263-267.
5. Paul R. Ehrlich, *The Population Bomb* (New York: Ballantine Books, 1968), p. 133.
6. J. C. Hoekendijk, *The Church Inside Out,* ed. L. A. Hoedemaker and Pieter Tijmes, trans. Isaac C. Rottenberg (Philadelphia: The Westminster Press, 1966), p. 184.
7. Noel McInnis, "Is the United States a Planetary Disease?" Spaceship Earth Curriculum Project, Center for Integrative Curriculum Design, Kendall College, Evanston, Illinois.
8. Nigel Calder, *Eden Was No Garden* (New York: Holt, Rinehart & Winston, 1967), p. 29.
9. René Dubos, *So Human an Animal* (New York: Charles Scribner's Sons, 1968), p. xi.
10. "Junkdump," 16mm, color, sound, 15 minutes, produced by Communic, Fenton, Missouri.
11. Kenneth E. Boulding, "The Economics of the Coming

Spaceship Earth," *Environmental Quality in a Growing Economy:* Essays from the Sixth Resources for the Future Forum, ed. by Henry Farrett (Baltimore: John Hopkins Press, 1966)
12. Joseph Sittler, "The Care of the Earth," *The Care of the Earth and Other University Sermons* (Philadelphia: Fortress Press, 1964), p. 97.
13. August Heckscher, "Man Above Nature—An Essential Inquiry," *The Twentieth Century Fund,* Annual Report, 1966, p. 1.
14. Edward T. Hall, *The Hidden Dimension* (Garden City, New York: Doubleday & Co., 1966), p. 3.
15. René Dubos, "Can Man Survive Life in Big Cities," *U.S. News & World Report,* May 1, 1967, pp. 64-67.
16. John R. Platt, "Life Where Science Flows," *Environment and Change: The Next Fifty Years,* ed. William R. Ewald, Jr. (Bloomington: Indiana University Press, 1968), p. 77.
17. Pierre Bertaux, "The Future of Man," *Environment & Change: The Next Fifty Years,* ed. William R. Ewald, Jr. (Bloomington, Ind.: Indiana University Press, 1968), p. 19.

FOR FURTHER READING

Ardrey, Robert. *African Genesis.* New York: Dell Publishing Co., Laurel Ed., 1967.

Bernarde, Melvin A. *Our Precarious Habitat: An Integrated Approach to Understanding Man's Effect on His Environment.* New York: W. W. Norton & Co., 1970.

Bonhoeffer, Dietrich. *Creation and Fall.* New York: Macmillan, 1965.

Bonifazi, Conrad. *A Theology of Things.* Philadelphia: J. B. Lippincott, 1967.

Boulding, Kenneth E. *The Meaning of the Twentieth Century* New York: Harper & Row, Colophon Ed., 1965.

Brown, Harrison. *The Challenge of Man's Future.* New York: Viking Press, Compass Ed., 1956.

Calder, Nigel. *Eden Was No Garden: An Inquiry into the Environment of Man.* New York: Holt, Rinehart & Winston, 1967.

Calder, Ritchie. *Man and the Cosmos.* New York: New American Library, Mentor Books, 1969.

Carson, Rachel. *Silent Spring.* Greenwich, Conn.: Fawcett Crest Books, rev. ed., 1970.

Charter, S. P. R. *Man on Earth: A Preliminary Evaluation of the Ecology of Man.* Sausalito, Calif.: Contact Editions, 1967.

Commoner, Barry. *Science and Survival.* New York: The Viking Press, Compass Ed., 1967.

Cox, Harvey G. *On Not Leaving It to the Snake.* New York: Macmillan, 1967.

Crespy, Georges. *From Science to Theology: The Evolutionary Design of Teilhard de Chardin.*, trans. George H. Shriver. Nashville: Abingdon Press, 1968.

Darling, F. Fraser and Milton, John P., eds. *Future Environments of North America.* Garden City, N.Y.: The Natural History Press, 1966.

Dobzhansky, Theodosius. *The Biology of Ultimate Concern.* New York: Meridian Books.

Dubos, René. *So Human an Animal.* New York: Charles Scribner's Sons, Lyceum Ed., 1970.

Edberg, Rolf. *On the Shred of a Cloud.* Trans. Sven Ahman. University Alabama: University of Alabama Press, 1969.

Ehrlich, Paul R. *The Population Bomb.* New York: Ballantine Books, 1968.

——— and Anne H. *Population/Resources/Environment: Issues in Human Ecology.* San Francisco: W. H. Freeman and Co., 1970.

Eiseley, Loren C. *The Immense Journey.* New York: Random House, Vintage Books, 1957.

Elder, Frederick. *Crisis in Eden: A Religious Study of Man and Environment.* Nashville: Abingdon Press, 1970.

Ewald, William R., Jr., ed. *Environment for Man: The Next Fifty Years.* Bloomington: Indiana University Press, 1967.

Fromm, Erich. *You Shall Be as Gods*. New York: Fawcett World Library, 1969.
Gilkey, Langdon. *Maker of Heaven and Earth: A Study of the Christian Doctrine of Creation*. New York: Doubleday & Co., 1959.
Graham, Frank, Jr. *Since Silent Spring*. Boston: Houghton Mifflin Co., 1970.
Hall, Edward T. *The Hidden Dimension*. Garden City, N.Y.: Doubleday & Co., Anchor Books, 1969.
Klausner, Samuel Z. *The Annals: Society and Its Physical Environment*. Philadelphia: The American Academy of Political and Social Science, 1970, vol. 389, May 1970.
Leopold, Aldo. *A Sand County Almanac: With Other Essays on Conservation from Round River*. New York: Oxford University Press, Galaxy Books, 1968.
McHarg, Ian L. *Design with Nature*. Garden City, N.Y.: The Natural History Press, 1969.
Morris, Desmond. *The Human Zoo*. New York: Dell Publishing Co., 1970.
Nash, Roderick. *Wilderness and the American Mind*. New Haven: Yale University Press, 1967.
Nicholson, Max. *The Environmental Revolution*. New York: McGraw-Hill, 1970.
Osborn, Fairfield. *Our Plundered Planet*. New York: Pyramid Publications, 1970.
Reinow, Robert and Reinow, Leona Train. *Moment in the Sun: A Report on the Deteriorating Quality of the American Environment*. New York: Ballantine Books, 1969.
Shepard, Paul and David McKinley, eds. *The Subversive Science: Essays Toward an Ecology of Man*. New York: Houghton-Mifflin, 1968
Sittler, Joseph A. *The Care of the Earth and Other University Sermons*. Philadelphia: Fortress Press, 1964.
Smithsonian Institute Annual Symposium. *The Fitness of Man's Environment*. New York: Harper & Row, 1968.
Storer, John H. *Man in the Web of Life*. New York: The New American Library, Signet Book, 1968.
Teilhard de Chardin, Pierre. *Building the Earth and the Psychological Conditions of Human Unification*. New York: Avon Book Division, The Hearst Corp., 1969.
_______. *Man's Place in Nature*. Trans. by René Hague. New York: Harper & Row, 1966.
_______. *The Phenomenon of Man*. Trans. by Bernard Wall. New York: Harper, Torchbook ed., 1965.
Udall, Stewart L., *The Quiet Crisis*. New York: Avon Books, 1970.
Whyte, William H. *The Last Landscape*. New York: Doubleday & Co., 1968.
Williams, Donald A. and others. *Christians and the Good Earth*. Alexandria, Virginia: The Faith-Man-Nature Group, 1967.